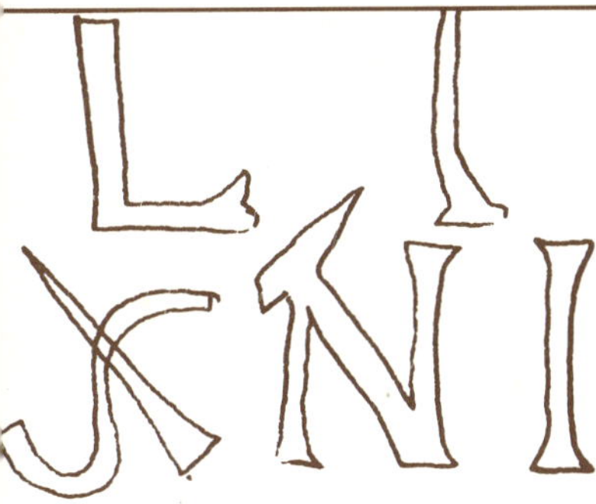

1632

THREE ROX WRECK
INDIANS TOLD FIRST SETTLERS TALE OF WINGED SHIP WRECKED NEAR HERE LONG AGO, OF STRANGE MEN AND BURIED TREASURE. WITH FINDING OF SOME MYSTERIOUS WRECKAGE, AND AN EIGHT-FOOT TALL NEGRO SKELETON DUG OUT OF A SHELL MOUND, SOME CREDENCE HAS BEEN GIVEN STORY.

TREASURE HUNTING NORTHWEST

BOOKS BY RUBY EL HULT

LOST MINES AND TREASURES OF THE PACIFIC NORTHWEST

Long-accepted treasure stories of Oregon, Washington, and Idaho are thoroughly investigated.

TREASURE HUNTING NORTHWEST

A sequel to Lost Mines and Treasures, *along with many new tales for treasure hunters.*

UNTAMED OLYMPICS: The Story of a Peninsula

True account of one of the wildest, strangest, most fascinating wilderness regions in America.

NORTHWEST DISASTER: Fire and Avalanche

First complete narrative of two of the worst disasters ever to occur in the Pacific Northwest.

STEAMBOATS IN THE TIMBER

Early steamboat days in Idaho, with many tales of the colorful characters who manned the boats.

TREASURE HUNTING NORTHWEST

BY

RUBY EL HULT

BINFORDS & MORT, *Publishers*

Portland • Oregon • 97242

FOR MAC

Treasure Hunting Northwest

ISBN: 0-8323-0087-X

Printed in the United States of America

FIRST EDITION

LIBRARY OF CONGRESS CATALOG CARD NUMBER: 70-165220

CONTENTS

PART ONE
RICH GULCHES

PART TWO
TREASURE TALES TWICE TOLD

A NOTE TO THE READER

Out of the enthusiasms of lost mine hunters and treasure seekers this book has been shaped — a fact amply evident in many parts of the text.

When my first lost mines and treasures book was launched, readers soon began to write me. Often it was to wish that some favorite story of theirs had been included: "I am sorry you did not tell about the Red Blanket Mine — I can't find out enough about it."

Sometimes it was to talk of lost mines and buried treasures whose stories had never been publicized.

Often it was to add new facts to the treasure tales I had presented, or to point out errors and discrepancies in the stories — some of the errors being mine, some of the discrepancies being history's.

Often readers wrote to recite their own adventures in treasure hunting — what theories they had followed, what conclusions they had arrived at.

Frequently they included clippings of lost mine and treasure stories from local newspapers.

Treasure Hunting Northwest grew out of research sparked by suggestions in the many letters and out of the happy wealth of the letters themselves. The book is a direct response to the avid interest which has been like a ground swell from Oregon to Missouri to New York, and from California to Texas and to Florida.

Let me repeat that I can supply no more information about the lost mines and treasures included in this book; I have told all that I know.

And let me again extend an invitation to all those who have other tales to tell or wish to comment on the stories in this book to write to me in care of Binfords & Mort, Publishers, Portland, Oregon 97242.

RUBY EL HULT

PART ONE
RICH GULCHES

Southwest Oregon

LOST VAULT OF GOLD

Jacksonville! That was a name to conjure with when the mining fields of the west were young.

Of all the Pacific Northwest towns which sprang out of the golden earth it was the one which most clearly conveyed the sense of brazen romance felt by those in hot pursuit of the golden goddess.

J'ville—so it was known by reasons of affection and brevity in its boom days of milling men and willing women, of gambled gold, of whiskey which tangled both foot and tongue.

Around it like spokes around the hub of a wheel turned all the mining excitements, discoveries, disappointments, developments, myths and mysteries of southwest Oregon . . .

Its story begins in 1851 with four packers who in charge of a loaded mule train left the fertile Willamette Valley and took the well-established trail which led southward to the settlements of gold rush California. After crossing the Rogue River they decided to stop to rest their animals along one of its beautiful side valleys. Partly this was a good excuse. What they were really interested in was doing some exploring.

In the Siskiyou Mountains, which bordered the Rogue River Mountains to the south, gold had already been discovered. The Northern Mines, as they were called in California, were producing richly if not yet so fabulously as the southern mother lode. On previous trips the packers had noted how closely the geological formation of the Rogue River area resembled that of the Siskiyou Mountains. They had heard too of tentative scratchings made along the Rogue by former travelers with encouraging though inconclusive results.

So while their mules cropped the lush grass of the valley the men spent their time prospecting. Late in 1851 near Daisy Creek they made what they realized was a phenomenal discovery. Each of the four, James Cluggage, James R. Poole, James Skinner and James Wilson, staked claims. The gold was so concentrated they could skim off hundreds of dollars a day. Hence the name Rich Gulch.

The group forgot their pack trip to the California settlements and spent an isolated but happy winter at their diggings. But by February 1852, having extracted the last of the food and ammunition from the mule packs, they found it necessary to send on of their number to Yreka for provisions. Which one of the four Jameses made the trip is not known, only that in Yreka when he paid for his purchases with the gold from the Oregon gulch it was noted by the merchants to be of a different color and of coarser grain than anything dug locally. Excitedly they asked him where he had got it but he evaded their questions skillfully.

That night Packer Jim got drunk in a saloon, spending more of the gold and saying there was plenty more where that came from. Yreka miners who heard this brag decided the visitor bore watching and they set up a system of surveillance. So a night or two later when the Oregon emissary thought he was slipping quietly out of town he was followed by a full half-hundred men all stealing silently one behind the other. They dogged him and his mule all the way back to the diggings on Rich Gulch, where they staked claims up and down Daisy Creek and in the surrounding gulches. As word of the big strike spread, more and more men stampeded in from California until a thousand men tramped the hills.

Promptly on the scene were two enterprising Yreka packers, Appler and Kenny, who put up a rude shelter on the valley's edge near the backdrop of timbered mountains and began to sell mining tools, tobacco and the so-called tanglefoot—a whiskey whose punch came not so much from alcohol as from cayenne pepper, plug tobacco and other imaginative additives. Tents and makeshift structures went up around that first place of business and within a month a log cabin was built.

Some of the early arrivals thought that in consideration of the identical given names of the four original discoverers the settlement should be called Jimville. Instead it took its name from

Jackson Creek on the north side of town—and Jacksonville was born.

Its progress was rapid. In 1852 Dr. Jesse Robinson opened a hotel, Union House. A dentist and a barber came and set up shop. A bakery which sold whiskey opened its doors. A Catholic mass was said. A Methodist church began building. Photographer Peter Britt came with a small wooden camera and established a studio. And in January 1853 before Jacksonville was one year old it became the seat of newly-created Jackson County.

All this activity was directly tied to the pounds of gold coming out of the nearby diggings. Two partners with a claim on the left fork took out more than $2500 in one day. One nugget alone weighed out at $1800. Of the original discoverers, James Skinner returned east with a fortune of $30,000 in gold dust.

And such tales spreading outward in concentric rings kept bringing new influxes to the Rogue River area of miners intent upon winning the same kind of easy fortune.

In the spring of 1853 a pair of determined adventurers arrived in Jacksonville whose names were James and Henry Wilson. They were brothers, not related to the James Wilson who had shared the Rich Gulch strike but the identity of name no doubt deepened their resolves toward sure success.

Most of the late arrivals (the spring of 1853 being late in the first Jacksonville boom) contented themselves with locating along the edges of the already proven local field: a little farther up the creeks, a little higher up the gulches. The Wilson brothers too explored the surrounding terrain but when they found no unstaked ground promising the returns they had envisioned they immediately developed restless ideas of looking farther afield. For in so new a country it ought to be simple to go out into the as yet unexplored hills and make a fresh discovery just as the packers had: find virgin ground where they could ladle off the gold like heavy, compacted cream. Why should they stay here and content themselves with blue skim milk?

They decided to strike out into the wilderness to the south on a prolonged prospecting trip. But in Jacksonville everyone tried to dissuade them from their plan. For the Rogue River Indians were acting up as usual and it was definitely unsafe to go into the mountains.

To the whites the Rogues couldn't have been better named. For that tribe had been causing trouble ever since travelers started passing through the country. As early as the spring of 1850 they had killed all but one of a party of California miners. In 1851 Governor Joe Lane had made a treaty with the main Rogue chieftain but the peace had soon been broken. In 1852 full-scale Indian war had raged with much fighting at nearby Table Rock. Now there were rumors that seven white miners had been killed at Table Rock and volunteer forces were preparing to leave from Jacksonville to go wipe out the Indian band.

The Wilson brothers remained unimpressed by these arguments. In the first place they were heading south, well away from the Table Rock trouble. In the second they did not believe any roving bands of redskins with bows and arrows and an outmoded gun or two could be much of a match for well-armed, straight-shooting white men. And they were able to round up a party of other men in Jacksonville who shared their views. Some were newcomers like themselves, disgruntled by local prospects; some were tough old prospectors afraid of little and ready for new adventure. Leading these men the Wilson brothers set off south on their prospecting trip.

The party had no immediate luck in finding any promising mineral ground but to their chagrin and alarm they ran into an Indian war party and were attacked. In the resulting skirmish one miner was killed.

This experience put the tenderfeet into a panic and even unnerved the hardy prospectors. For the warnings they had so blithely ignored in Jacksonville now took on sinister reality. The men decided the better part of valor would be to return to Jacksonville until the Indians had been pacified. All except James and Henry Wilson, that is, who were at once more daring and reckless, perhaps more foolish, and certainly more tortured by the itch for fortune. When they insisted on continuing their quest alone the others gave them most of the remaining provisions and ammunition.

The Wilson brothers, feeling a certain amount of inward contempt for these friends who were scuttling off toward safety, waved them goodbye, saying: "Good luck to you. We hope you have no trouble on the way back to Jacksonville. But we can't go back. We know there's gold somewhere behind that smoky line of mountains"—pointing to the range that barred the way southeast. "We're going to dig it out no matter how many redskins guard it. We'll come back rich as kings, boys. Either that or leave our bones to bleach here in the wilds."

The place of parting was beside a mineral spring at the base of a tall rock chiseled by the elements to resemble, in rough form, the two segments of an hour-glass. On departing from this natural landmark their real search began.

With provisions and supplies loaded on two pack horses they had retained, they rode resolutely through the range they had indicated to their departing friends, always keeping a vigilant watch for roving Indians. The way was rough with only game trails to guide them, but fortunately they encountered no natives.

At length they worked their way into a "green and narrow valley, walled in by precipitous mountains, around which meandered, over and among the boulders of richly colored rocks and across beds of smooth and shining pebbles, the limpid waters of a snow-fed stream." The valley was a natural retreat, peaceful and safe, and since the men were tired from their travels and their animals showed signs of wear they decided to remain in the place to rest.

In need of shelter not only for themselves but for their supplies, saddles and other gear, they thought it best to build a simple log cabin, leaving fort-like loopholes for defense in case they should be attacked by Indians.

Early next morning Henry set out to cut timbers for the cabin. As they were in great need of fresh meat James took his gun and crossed the valley looking for game. He killed a deer, partially dressed it out, then hoisted the carcass to his shoulders and started back to camp. By now the sun was high in the sky, its rays hot. He was so thirsty when he reached the stream in the valley he threw his heavy burden on the grassy bank and knelt on a sandbar to drink. As his lips touched the water his eyes were drawn to the sparkle of bright pebbles along the creek bed. Plunging his hand into the water he scooped up a handful of gravel and examined it.

"It was gold!

"Yes, there it was, in coarse yellow grains and lumps—richer than a dream." He had stumbled onto wealth which was "sudden, splendid, exhaustless," and he and his brother could now say goodbye to the poverty of the past.

His shout of wild joy bought Henry running from the other side of the small valley. When Henry understood what James had found the two did a triumphant dance. Even as they had boasted, they would return to Jacksonville "rich as kings."

In a state of feverish excitement they hurried to finish the rude cabin and as soon as everything was under cover they returned to the creek to begin placer mining. The gold had already been panned and winnowed by nature so that it was easy to extract it from the stream bed. Within two weeks they had accumulated a large golden heap.

The brothers had a talk and decided it was best not to leave the valley yet. While they had been prospecting with the men who had returned to Jacksonville they had heard from settlers that Joe Lane, now Superintendent of Indian Affairs for Oregon, was coming to try to restore peace along the Rogue. That would be a difficult process at best, so the longer they could remain in the hidden valley the more apt the roads and trails were to be safe when they ventured out. It would be wise to bury the gold so if Indians should surprise them here they could flee unencumbered. Then if they escaped they could return for it later.

In the center of the cabin floor they dug a large hole and lined it with closely-fitted rocks. Into this vault they deposited deerskin bag after deerskin bag of gold, for during their period of waiting they continued to work at the creek and to secure wondrous amounts of beautiful, pure dust and nuggets.

They could not possibly spend the winter in the valley for they had not enough supplies or ammunition. When the leaves began to turn they knew that, safe or not, they must leave their refuge and make their way back to Jacksonville. Probably Joe Lane had the Indians calmed down by now.

What they could not know was that although Lane had come south and had managed to sign a treaty with the Indians on September 8, 1853, this peace did not last either. In October two

Jacksonville merchants were killed, and there were other killings along the river. Bands of murderous Indians, some from the Shasta tribe, roamed the hills. On the very morning that James and Henry Wilson prepared to leave the valley one such Indian war party slipped stealthily into their hideout.

The brothers went out early to round up their saddle and pack horses, which they found looking sleek and frisky. As they returned to the cabin Henry was in front leading two horses, James in the rear with the other two. Suddenly a whoop rang out, followed by rifle fire. Henry fell, apparently dead, as the horses began to scream and rear, and James saw a number of Shasta braves burst from the woods and run toward him. Quickly he raised his rifle and fired, causing them to retreat to the woods. On emptying his rifle he pulled out his revolver, grabbed the halter of the nearest horse, leaped to its back and fled down the valley. Sure that there was no hope for his brother he kept going and after a ride of a few hours realized he had made his escape. He saw no more of the Indians.

After a nightmare trip he reached a settlement on the California side of the line. He was safe but the strain, shock, hardship and exposure he had endured left him weak and ill. His health seemed broken. Much in need of medical attention he did not return to Jacksonville but went instead to San Francisco. There he consulted various doctors but failed to improve. Bitterly he missed his brother Henry, from whom he had always been inseparable. Sick, sad and lonely, he began to write to a cousin of his, Ted Harper in Chicago. In long, rambling letters he told his cousin the whole story of the prospecting trip out from Jacksonville, of the valley where he and Henry had holed up, of his discovery of the gold, of their mining, of the burial of their hoard in the rock-lined vault and of that fatal morning when Henry was killed. As soon as he was well, he wrote, he was going back to the valley and dig up the cache of gold. He had no doubt that he could find the location.

But in the fall of 1859 James Wilson died in San Francisco. Ted Harper, on being notified, traveled to the Bay City to settle his cousin's affairs. Among the papers found in Wilson's boarding house room Harper discovered one last unfinished letter:

> San Francisco, Oct. 26, '59
>
> Dear Cousin: I had hoped to see you before this, but the end has come sooner than I expected. I dreamed of that cabin in the wilds of Oregon, last night, and saw poor Harry fall again before the cowardly shots of those cowardly Shastas; and I think it is nearly over. I must write what I intended to have spoken, and endeavor to give you such directions as will enable you to find the cabin, for you must find it, Theodore, and enjoy its hidden gold. The first part of your course is plain enough: Start from Jacksonville and keep the California road for—

and there it broke off, blotted with ink as if the writer might have been seized suddenly with illness and dropped the pen.

The Proposition

It was a rainy April night some years later. Sam L. Simpson sat in the tiny cubicle of his lately opened law office in Portland, Oregon. He was having a fit of the blues. With the ink barely dry on his license to practice he had come to the Oregon metropolis believing it the place where ample cases with fat fees would be fastest arriving at his door. Instead the city had proved to be overrun with lawyers, with not enough acrimony and litigation among the populace to support them. No cases large or small had come to him and his scant funds were rapidly disappearing. What was he to do, he wondered disconsolately, when his money should run completely out.

Hearing a tap at his door he gave a gruff "Come in." A tall, dark young man dressed in a brown suit entered. Sam Simpson recognized him as Ted Harper, a fellow roomer in the boarding house where he lived.

After some banal remarks about the nasty night out, Harper focused keen eyes upon the young lawyer and said he had come to talk business. Simpson waved him to a chair.

"Are you satisfied with all this?" Harper asked, indicating the small confines of the shabby office and the one shelf of heavy law books. "If you succeed at all in the practice of law it will be only after years of patient grinding and penny-pinching. Tonight I have come to share a valuable secret with you and to get your help in a project which, if successful, will make us both rich."

Simpson, sure he was to be made a shady proposition, gave Harper an impatient, contemptuous stare.

"No, no, this is a perfectly honorable matter. If you'll promise not to divulge what I disclose, in case you do not join me, I will tell you the story. Did you ever hear of the Lost Cabin?"

Simpson gave a pledge of secrecy and Harper launched into a recital of the adventures of his cousins, the Wilson brothers, climaxing the tale with mention of the many buckskin bags of gold cached beneath the earthen floor of the Lost Cabin in the hidden valley. And taking a worn sheaf of letters from his pocket he extracted one missive and handed it to Simpson.

It was the last unfinished message of James Wilson to Harper. Reading it Simpson felt a prickle along his neck as the reality of the lost vault of gold was borne in on him.

"That's really all I know," Harper admitted as he put the letter away. "Maybe it isn't enough to go on. But here's my idea. Let's purchase an outfit together and spend the summer in the mountains of southern Oregon. With what clues we have, I believe we can find the cabin."

Simpson sat in thought. What did he have to lose? he asked himself bitterly. Anything would be better than to sit all summer in this office, embarrassed by a lack of clients and giving away more and more to panic as he watched his money run out. And if only they could find the hoard of gold, then his troubles would be over.

"All right. I'm with you." The two men shook hands.

Relief was all that Sam Simpson felt the day he packed away his law books and returned the office key to the landlord. He and Harper then bought themselves saddle horses. They stocked up on provisions to last for several months, added some tools and other equipment, and acquired two pack horses to carry their gear.

In mid-May when the snow should be fairly well gone from the mountains they started south. Their first stop was Jacksonville, the point of departure for the Wilson brothers more than a dozen years ago. They found the Methodist church finished, a sawmill established, a theater open. Added were two breweries, a city newspaper, A. A. Beekman's popular express office. and a new city hall with jail cells. The Table Rock Saloon and other groggeries were flourishing. Streets had even been graded and graveled. A hearty optimism prevailed among the town populace.

Rich Gulch and other mining properties had proved unfortunately shallow and quickly worked out but other Rogue River placer strikes had followed in localities close enough that Jacksonville could serve as supply center and rousting place. Then the discovery in 1860 of the wide Gold Hill ledge twelve miles northwest revealed the potential of quartz-mine riches.

Simpson and Harper, listening with sharps ears to this recital of local history, admitted happily that the Wilson brothers had been right. In 1853, the year of their quest, much of southwest Oregon had lain virtually unexplored and unprospected: the hills were indeed a mecca of undiscovered Rich Gulches. The Sterling Creek strike of 1854 had yielded brief staggering amounts of gold like those James Wilson claimed for the hidden valley. Althouse Creek, Sucker Creek, Applegate River, Coffee Creek — all were names which stirred memories of notable strikes. So there was every reason to believe the Wilson account true.

With a new elation and a stronger sense of conviction the two treasure hunters started on south again down the stage road, keeping to it for one day, then turning off into the wilds southeast. The going was slow and tortuous as they pushed through heavy brush, clambered over fallen timber, passed unknown lakes and through somber solitudes. As Simpson said: "What a road to fortune!"

Their first encouragement came one day when they reached the base of a rugged chain of mountains trending westward and directly in their course. "A few miles to our left we found an accessible path; but, surveying the range with our glass for a great distance on either side, we could see no other depression in its clear-cut line. This, then, was the only gateway to the southeastern country beyond, and through it the Wilson party must have passed, without a doubt. At the entrance of the pass we found the dim vestiges of a campfire and began to hope we were right. Farther on we found where an axe had been used here and there and knew that we were on the footsteps of white men. Was it the Wilson trail? . . .

"The western peaks were crowned with sunset gold, and our day's march was almost done, when unexpectedly we haulted before a lofty boulder shaped like an hour-glass; and at its base, staining the rock, over which it flowed, we found a spring—bitter with mineral constituents."

Harper and Simpson were pleased, feeling the battle half won. For surely this was the important landmark they sought: the place where the Wilsons had parted from the others of the prospecting party; the place from which they had set out alone to attack the mountain barrier which lay so close.

After resting two days the young treasure seekers rode on, still southeast through the mountain pass, into wilder, more rugged country—a "region of volcanic agonies, of fierce upheavals and lava floods." But they failed to find the hidden valley where they believed it to be. Both became morose, touchy and tired. For his part Simpson was so worked up he began to experience a kind of disorientation. All night in his dreams he would repeat endlessly the troubles and trials of the day and during the daylight hours would hardly know whether he was waking or sleeping.

One day about sunset they made camp. Harper, exhausted, fell asleep with his head on a roll of blankets, while Simpson sat with his back to a tree trunk and with a pencil idly in hand reread for the hundredth time James Wilson's maddeningly broken-off letter. If only there were some way to know how the instructions would have gone on from there! No sooner had this thought gone through his mind than looking up he saw before him the ghostly figure of a miner: a tall, heavy-set man with a bushy beard, dressed in a gray flannel shirt, without coat and with a revolver strapped at his side. As the stranger cast sad and sympathetic eyes upon him Simpson seemed hypnotized.

Then he heard Ted Harper up kindling the fire and roused himself. The letter he had been reading had slipped from his fingers, and as he stooped to retrieve it he noticed in the space just below the ink blot some pencillings he had never seen before. The drawings seemed to represent two ranges of mountains meeting at right angles, and below that a miner's pick.

Was he losing his mind? Sam Simpson asked himself. Had he, while dozing off, made the drawings himself? Or had the ghost of James or Henry Wilson visited him, attempting to give directions to the Lost Cabin?

In doubt of his own reason, Simpson said nothing to Harper. But two mornings later when they climbed a high peak to look over the surrounding country for clues as to where next they should go, Simpson saw another range of mountains intersecting at right angles the range through which they were passing. A light shone in his mind: The miner's pick, a symbol of gold, just below the intersection of the two ranges—that must mean that the valley they sought lay where the two ranges met.

With confidence in his sanity restored he confided in Ted Harper the occurrence he still could not explain. Both felt stirred and excited.

From their high vantage point they saw their way lying clear before them, and they traveled down, following a path that led them mysteriously along.

"On—on we went in a dream of wonder and future wealth, and nothing impeded our progress now, until at last we entered a narrow valley walled in by precipitous mountains and bordered on each side by a beautiful stream. We knew we were upon sacred ground; and along the shadowy fringe of the forest, where the fretted waters sang a barbaric rune, we rode, silent as spectres. A resistless magnetism drew us on, and not a word was spoken . . . . We turned a projecting angle of the wood, and a square, black object half buried in a tangle of weeds, was before us. We had found the Lost Cabin!—nothing now but an empty pen of scorched and blackened logs."

Hastening to dismount they extracted their tools from the pack horse loads and entered the enclosure—only to stop abruptly as they saw a human skull leering at them. Poor Henry Wilson—yes, it must be his remains. The Indians had no doubt dragged his body inside and then set the cabin afire in an attempt to cover up their murderous deed. But only the roof had been completely consumed; the lower logs had been too green to do anything but smoulder.

The sight of the skull saddened them but they were not to be deterred from their long-delayed treasure dig. Sam Simpson raised his pick and sunk it into the center of the cabin floor. Again and again he drove it through the loose soil. The point of the steel struck something hard—no doubt one of the rocks which enclosed the vault.

At that instant a shot rang out and Ted Harper fell at his feet with a terrible cry. Later on, when Simpson was able to reflect on the tragic event and reconstruct the sequence of happenings in his mind, he concluded that Harper had picked up one of their guns to move it out of the way and had accidentally discharged it, the bullet striking him fatally. But at the time, with Simpson already overtired, with nerves strung too taut from the suspense of their quest, the shock proved unendurable. He went completely out of his mind and remembered nothing more. When he came to himself, much later, he was back in Portland, recuperating from an illness of both mind and body. He was told that he had been picked up by a stage driver who had found him wandering along the road near the California border, a nude and gibbering wildman. His only impression of the interlude was of "night, a long starless and dreamless night of clouded intellect and slumbering soul." He must have rushed madly from the Lost Cabin and through the wilderness, and being reduced to his animal instincts had somehow found food and a way to safety.

The Life of the Poet

Or at least this is the story as Sam Simpson wrote it many years later. His tale, entitled "The Lost Cabin," was published in the historical magazine, *Oregon Native Son,* September 1900, a few months after the author's death.

Is it fact or fiction? Since the *Oregon Native Son* did not publish fiction it must have passed for true experience. Even so, how much of it can we believe?

Samuel Leonidas Simpson was a real person, and a review of his life provides some enlightening information, as well as some evidence—purely circumstantial—that the treasure search took place.

Born in 1845 Sam was brought by his parents to Oregon as a year-old child and he grew up in a cabin in Marion County.

He proved to be a poet both in talent and temperament. his enchantment beginning as a youth when someone presented him with a book of Byron's works. He was soon writing creditable verse of his own.

But because the craft of poetry could not offer him much in the way of monetary return he was forced to make compromises with his muse and seek a way to scratch out a living. After graduating from Willamette University at Salem in 1865 he decided to read law and was admitted to the bar either late in 1867 or early 1868. His sensitive soul found it hard to adjust to the world as it was, and he hated law from the beginning. He began to solace himself with drink and ended an alcoholic. As was written of him later when his verses had brought him fame, he was "the most drunken poet, and the most poetical drunkard that ever made the muses smile or weep."

It is true, as he says in "The Lost Cabin," that he set up an office in Portland right after being admitted to the bar and that his clients were few or even nil. According to Ray H. Glassley he closed his office on April 11, 1868. This fits with Simpson's statement that Harper entered his office on an April night and he packed away his law books a few days later.

Simpson tells that he and Ted Harper waited until mid-May to go south so that the mountains would be free of snow. This would have given him nearly a month of leisure. And for a poet, what better use of time than to let emotions crystallize on paper? Again according to Glassley, a week after he closed his law office Simpson wrote his most beloved poem, "Beautiful Willamette":

Onward ever
Lovely river
Softly calling to the sea.
Time that scars us
Maims and mars us
Leaves no track or trench on thee.

That he was already aware of his own tragic flaw is reflected in the poem. It was his drinking which scarred him, maimed and marred him; and he no doubt accepted Ted Harper's proposition partly because of a need to flee from himself and temptation. And since he disliked the practice of law and wished only to be free to write poetry, his lively imagination must have leaped joyfully at the thought of finding enough gold in the Wilson cache to enable him to chuck his law career and develop his artistic talents.

So the story could be true. But the fantastic happenings described form a large stumbling block to complete belief.

Would Ted Harper have revealed the details of the Wilson cache to a comparative stranger, asking him to join the search? It is my thought that perhaps it didn't happen just as Simpson tell it. Since he and Harper lived ot the same boarding house they may have done some drinking together. In some bibulous exchange Ted may have confided the story of his cousins' adventures, with Sam's impulsiveness and enthusiasm putting the flame to the search. If the final agreement regarding the treasure hunt was reached in the lawyer's office then the tale would be virtually true as related but with some unpalatable sidelights conveniently forgotten and events rearranged so as to heighten the drama.

And what are we to make of the apparition of the miner appearing to Simpson, leaving behind the pencilled drawings with clues to the location of the hidden valley? Those who believe in ghosts will have no trouble here, but I for one wonder how much liquor Harper and Simpson had with them. Simpson's description of his disorientation, his loss of touch with reality, sounds like a man in a nervously drunken stupor, not too far from the D.T.'s. In which case both the drawings and the subsequent inspiration as to where to look for the valley may have come from the bottle. And if he and Harper were drinking then the accidental shooting which ends the tale becomes all too convincing.

As for Simpson's later life he married Julia Humphrey by whom he had two sons. He practiced law in Corvallis for two years, then went into journalism, a profession more suited to his tastes and talents. He continued to write poems which were widely published and so well received in his home state that he became known as "Oregon's Sweetest Singer." But his days were still so riddled by drink that those who knew him best adjudged his life a failure. Only fifty-five when he died he is buried in

Portland's old Lone Fir Cemetery where a memorial marker pays tribute to his achievements in poetry.

My own conclusion is that his story "The Lost Cabin" is based on fact but represents a certain amount of author's license in concealing what it was not expedient to reveal and in shaping the episode into a dramatic narrative.

If the story is true, at least in broad outline, where then lies the valley with the buried buckskin bags of gold? Admittedly the clues are irritatingly vague and frustratingly sparse.

The starting place for a search is, of course, Jacksonville. Other mining centers boomed and vanished but J'ville somehow eluded the ghost town phase. Instead it has become a living memorial to the past—a place where history is lovingly and sensibly preserved and restored. In its handsome old Museum a search through old letters, diaries, newspaper files and other records might reveal more facts about the Wilson brothers and their prospecting trip, more details of the Simpson-Harper trip and its outcome.

Failing that, Jacksonville is still the starting place. For no better atmosphere exists for projecting the treasure seeker back into those first free prospecting days when, if one could elude the Indians, the gold fields lay in the hills for the finding.

From J'ville the trail leads south along the old stage road for one day's trip by saddle horse. The direction is then southeast into country that is still rough and wild.

Perhaps it would not be too difficult to find that first significant landmark: the hour-glass rock towering over the mineral spring. This place should lie at the edge of a range whose most distinctive pass leads still southeast. But where then? To the intersection of another range of mountains. But what range? In what direction? Clear to the Skyline Trail? Or down toward the California Siskiyous? It is to be remembered that both James Wilson and Sam Simpson, after their different flights, found themselves on or below the California border.

And then, which way to the hidden valley? If only it could be found it might well be recognized from its description. And possibly some vestiges of the ruined log cabin may still remain —guarded by two eyeless and grinning skulls.

INDIAN LOST CINNABAR MINE

Waldo Jewett, Robert Spencer Jewett and Thomas Drake Jewett were all born in Athens, Ohio, the sons of Leonard and Emma Jewett. The three brothers had grown to adventurous young manhood before the family came west over the plains in 1850. The parents took a donation claim on Sauvies Island in the Columbia River northwest of Portland, Oregon, but the boys left almost immediately for the gold fields of California. Where they mined in the mother lode country and what their experiences we do not know, only that they returned to Oregon in 1853 with pokes full of gold.

It may well have been the Rogue River excitement and the cry of Jacksonville that lured them back. They found the trail through Yreka and on north thick with stampeding miners. Even en route between the Willamette Valley and Jacksonville, where the Rogue River crossing must be made, travel was heavy. The river crossing had been an early major obstacle along the Oregon-California trail but with the Rich Gulch strike a man of somewhat unsavory reputation, Davis Evans, took a north bank claim at the mouth of the creek which became known as Evans Creek and put a ferry across the river. He did a rushing trade but had turned restless and was ready to move on by the time the Jewett brothers came along with their California-won riches. The Jewetts thought they saw a good opportunity at the site and bought Evans out, both north bank claim and ferry.

All the commotion of the times centered around Jacksonville but as the mining town lay some fifteen miles beyond the Rogue crossing travelers frequently wished to rest at the ferry before proceeding on. So the Jewetts found it expedient to add a store, to serve meals and to offer care and shelter for animals.

The same Indian troubles which kept Jacksonville in a state of alarm made life at the ferry dangerous. The Jewetts lost property during the Rogue River war which ended with Joe Lane's Indian treaty of September 1853. Outrages committed by unscrupulous whites caused the Indians to take the warpath again in 1855 with battles ranging back and forth across the Rogue, with many atrocities committed on both sides. During one battle in 1855 the ranch house at the Jewett ferry was fired on.

The Indians were led by Chief John, Rogue River George, Limpy, and chiefs of the Cow Creek and Galice Creek bands. So many Indians were killed that in June 1856 most of the bands agreed to go to a reservation. Only Chief John refused to surrender. "This is my country," he said. "I was in it when these large trees were very small, not higher than my head. My heart is sick with fighting, but I want to live in my own country . . . . I will not lay down my arms and go with you to the reserve."

He and his band engaged Captain A.J. Smith's forces in one last desperate battle in which one-third of the white soldiers were killed or wounded, the remainder being rescued only when Captain Augur arrived with reinforcements. After that John's war-weary people declined to fight further and he went with them to a coastal reserve.

Right after Chief John's last stand a brave who panicked at the idea of being confined to a reservation came fleeing to the Jewett ferry. This Indian, also known by the name of John, was taken in and befriended by the Jewett brothers, who allowed him to sleep in a shed, fed him and gave him odd jobs to do and made no objection when he went off on long trips into the hills where lay his former hunting grounds. As time passed John began to pine for his family and friends, and deciding that reservation life might not be so onerous after all said goodbye and took himself off in the direction of the coast.

During these years the Jewett brothers lost none of the verve which their California mining success had generated. Their ferry and roadhouse business came first but any time they could spare they spent prospecting in the hills and what with rich strikes being made all around them they held high hopes of making a strike of their own. Waldo Jewett died in 1859 but Robert and Thomas kept up their prospecting activities and in 1860 filed on a gold-quartz claim on the south side of the Rogue River, west into Josephine County. They named the mine the Elizabeth but it was usually referred to as the Jewett lode. In association with D. William Douthitt, a Jacksonville mining man who acquired a one-sixth share, they began work on the property.

During that same summer of 1860 a new mining contagion reached the ferry. It was a "bug" which had come to life during the California gold rush and now came traveling up the trail from San Francisco. Robert Jewett was immediately infected: he developed a fever for quicksilver.

In the California coast ranges Indians since time immemorial had visited certain hills of red earth to obtain pigments for their war paints. As early as 1827 the Spaniards had inspected these slopes and recognized the red color as deposits of cinnabar but they did not test the ore until 1845 and did no mining of consequence.

The California gold stampede, touched off by Marshall's discovery of nuggets in the millrace at Sutters Fort in 1848, brought on an intense need for mercury. An American company took over the Spanish diggings twelve miles southeast of San Jose naming the mine New Almaden after the richest quicksilver mines in the world at Almaden, Spain. So high in quicksilver yield was the cinnabar ore that the New Almaden soon became second in production only to the old Almaden.

In the early 1850s new California cinnabar deposits were discovered in the same general area, a little to the south on the slope of San Carlos Peak. The richest of these deposits was named the New Idria for the quicksilver mines of Idria, Austria. Both the New Almaden and the New Idria proved so tremendously wealthy that others beside their developers claimed them and the noise of acrimonious law cases thundered through the courts and the early-day newspapers. The litigation almost more than the rich dividends the mines were paying made the names New Almaden and New Idria household words throughout the west. Prospectors became aware that to discover a really good mercury mine might be better than to stumble upon a hill of gold.

Such was the background of a strange drama enacted at Jewetts' ferry—a drama giving rise to a lost cinnabar mine legend which still challenges treasure seekers across the land.*

On a July day a man came riding north up the road on horse-

---

* Proved by the fact that a request for information about this lost mine has come recently from a man in far-away Brooklyn, New York.

back (there were no stages through these parts as yet) and crossed over on the ferry. He was from San Francisco, he said, on his way to Portland. Having a long journey still before him he stopped at the ranch for dinner and to have his horse fed.

Over his years at the ferry Robert Jewett had built up a collection of unusual rocks, mineral specimens and curios of interest. Most of his pieces had been brought in from the hills by himself and his brothers but some few had been given him by friends and visitors. As the collection was kept in the public room the traveler, while waiting for his meal, walked around inspecting the different items. He came to a stop before one big piece of red rock, lifted and hefted it and examined it carefully. "Where did you get this?" he asked Robert Jewett with some intensity.

"What is it?" countered Jewett.

"You mean you don't know? It is a piece of cinnabar. Sulphide of mercury. The richest piece I've ever seen. If you've got a ledge of ore like that around here you're worth a fortune. Yes, worth a fortune."

Robert Jewett muttered he had forgotten just where the specimen had come from and the matter was dropped. But after the traveler had finished his meal and ridden on north the ferryman stood in turmoil, searching his memory. Who was it had given him the rock? Wasn't it Indian John? Yes, he remembered it now. John had brought it back from one of his trips into the hills and had handed it over without saying much. Robert had recognized it as an unusual specimen and so had placed it in his collection but he had never bothered to ask the Indian where it came from. Only John would know, and the last Jewett had heard of him he was living on the Siletz Indian Reservation. Maybe it was still possible to find him and ask him about the rock.

The first thing to do was to make sure that the piece of cinnabar was as rich as the passing stranger had claimed it to be. The specimen being a large one weighing about six pounds, Jewett broke it in two and took half of it to Jacksonville expecting to look up a friend and ask him where to get an assay made. But his friend was away on a trip to San Francisco.

The name of this man Jewett intended to consult in Jacksonville remains unknown but since he was a mining man who frequently traveled to San Francisco where he was well-acquainted in mining circles, he could have been D. William Douthitt who had invested money in the Jewetts' Elizabeth Mine. So for convenience sake we will refer to him as Mr. D.

Jewett asked another friend to box the specimen and take it to Beekman's for dispatch to the Bay City via Wells, Fargo Express. With it should go a note requesting Mr. D. to have an assay made during his stay in San Francisco.

When the specimen reached Mr. D. he took it to what he regarded as the best assaying firm in town, Justh, Hunter and Company, where he was personally acquainted with E. Justh, a Hungarian with a long history of mining experience in his own country and of topnotch assay work in San Francisco.

When Mr. D. returned in a few days' time for the assay report Justh began shooting questions at him: Where had the specimen been found? How extensive was the prospect? Who owned it? He then handed over the assay certificate, which gave, as Mr. D. remembered it later, "above 70, and I am sure it was a fraction less than 74 per cent." Justh commented that the specimen was the richest piece of cinnabar ore he had ever seen, much richer than any ever brought from the New Almaden Mine. He was willing to send an expert to Oregon to look over the new mine and if it proved as rich as the specimen indicated it might be he would gladly put up the money for opening and developing it.

Mr. D. was well aware that such an offer from a man of Justh's prominence and mining experience was impressive but he was forced to answer that a friend had forwarded the piece to him for assay and that was all he knew. He promised to find out the particulars about the mine as soon as he got back to Oregon and write Justh immediately.

So on returning to Jacksonville in a state of animation he was much let down to learn that Jewett, the owner of the richest piece of cinnabar ever found, hadn't the foggiest notion where it came from. The secret resided with an Indian on the Siletz Reservation.

Apparently Thomas Jewett was away from the ferry at the time. The facts about the rich specimen were shared only by

Robert Jewett, Mr. D. and the friend who had taken the package to Beekman's express. The three passed the assay back and forth among them, kept repeating Justh's words "richer than anything that has come from the New Almaden," and exclaiming over the offer of money to help open the mine. It was obvious that they were close to the biggest thing in their lives and they were determined to let no obstacles stand in their way in learning the whereabouts of the ledge from which the ore had come. The three banded together in a copartnership with the avowed purpose of getting the Indian back from the reservation so he could lead them to the place where he had found the piece of cinnabar. Since the Indian was a friend of Jewett's they felt sure of his cooperation.

It was their good fortune that Mr. D. was on friendly terms with the Reverend Edward A. Geary, Superintendent of Indian Affairs, whose office was in Portland. This suggested a course of action. Mr. D. would write to Geary requesting that the Indian be allowed to leave the reservation and visit his old friends at the ferry. The letter would be delivered by a special messenger in the person of Robert Jewett, since only he could be sure of recognizing the right Indian. On leaving for Portland he would take an extra saddle horse with him so if permission were granted he could come back by way of the Siletz Reservation and bring Indian John home with him.

Everything happened exactly as planned. Superintendent Geary acceded to the request and gave Jewett a written authorization to the Siletz agent. Jewett had no trouble in finding and recognizing his Rogue River Indian friend, who was happy to see his benefactor and more than willing to return to his old tribal grounds.

As they rode away from the reservation Jewett did not explain to John why he had come for him or what was wanted of him, thinking it best to wait until the Indian was again settled at the ferry, feeling contented and at home. Then he would question him casually, so as to avoid arousing either his suspicion or his greed.

Mr. D. and the other copartner were on hand at the ranch to welcome Jewett and Indian John and to congratulate Jewett in private on his success. Then they left to let the ferryman work his wiles on the Indian.

At dinner time a few days later when John came in to eat in the room where the mineral specimens were kept, Jewett took the remaining half of the cinnabar rock off the shelf and began to pick at it with his knife. The Indian showed no recognition of the piece. Then Jewett, thinking to arouse his curiosity by demonstrating to him the properties of the rock, put the piece in the fire where the dinner was cooking. But what John evinced was not interest but alarm. In a panic he jumped up and dashed out of the room.

Jewett of course realized he had made the wrong move and hastily raked the ore from the fire. He then ran out after his Indian friend, calling him back. John paused in his flight but would not come near the house until Jewett went to him, talked to him soothingly and showed him that the specimen was no longer in the fire. At length he came back inside but showed himself still wary of the stove.

That night Indian John refused to sleep in the shed which had been his special refuge; instead he took his blankets off into the woods to an old Indian encampment. In the next days he avoided Jewett, acting surly, suspicious and uncommunicative.

It was obvious to Robert Jewett that he had somehow touched upon the superstitious fear of the Indian — but what fear? what superstition? John probably had no words with which to explain it to a white man even if he wished, which he probably didn't. And while this overwhelming dread was on him, asking him questions was futile.

There was one tried-and-true way of unlocking an Indian's reticence, and in his desperate desire to learn the location of the cinnabar ledge Jewett began to ply his red friend with whiskey. Of course he knew it was illegal and immoral and twice reprehensible because he had made himself responsible for the Indian in taking him off the reservation. He found the Rogue Indian as fond of firewater as the next aborigine and happily cooperative in downing all the drinks given him, going from tipsy to reeling to dead drunk. Through all these stages Jewett kept questioning him and doing everything possible to worm out of him the secret of the ore. But he got no more satisfaction from John drunk than he had from John sober.

Robert Jewett withdrew the liquor and for some days kept a wary, watchful eye on his protege while he contemplated what move to make next.

After a few days the Indian seemed to throw off his anxiety. He became more talkative and friendly with Jewett, and one day called him down to the river indicating he wished to pow-wow. Slowly and haltingly he told the following story:

When he was very young, before the white men had come into the country, the chief enemy of his tribe were the Klamath Indians. Always before raiding the enemy the Rogue war parties painted themselves as wildly and gaudily as possible, the better to outshine their foes. So it made a big furor in the camp when a young brave came in from a deer hunt painted a vivid color of red none had seen before. On being questioned by the tribal chiefs the warrior said that about a day's chase away he had shot a deer and stopped to build a fire. Nearby he had noticed some odd-looking red rocks and while cooking his venison amused himself by pounding one of the rocks on another flat-topped rock, causing a fragment to fly into the fire. As the rock heated it turned to brilliant particles, which he later raked out of the fire. When the mass had cooled he had painted himself—the only trouble being that it gave him a slipperiness of the mouth and other peculiar sensations.

A paint-rock! Decked out in such a glorious color the Rogue River war parties would surely defeat their enemies. So said the chiefs as they ordered the women to pack up the camp so they could all follow the hunter back to his discovery.

After a day's journey they arrived at the ledge of red rock, where they turned out the horses, set up the camp and kindled a big fire. As the blaze died down they laid pieces of the colored rock in the coals and while waiting for the material to melt and to cool they feasted and danced.

When all was in readiness they indulged in a lavish painting of the braves and their mounts. More dancing and merriment followed until everyone was exhausted and lay down to sleep. In the morning the warriors would repaint themselves and set out on the warpath.

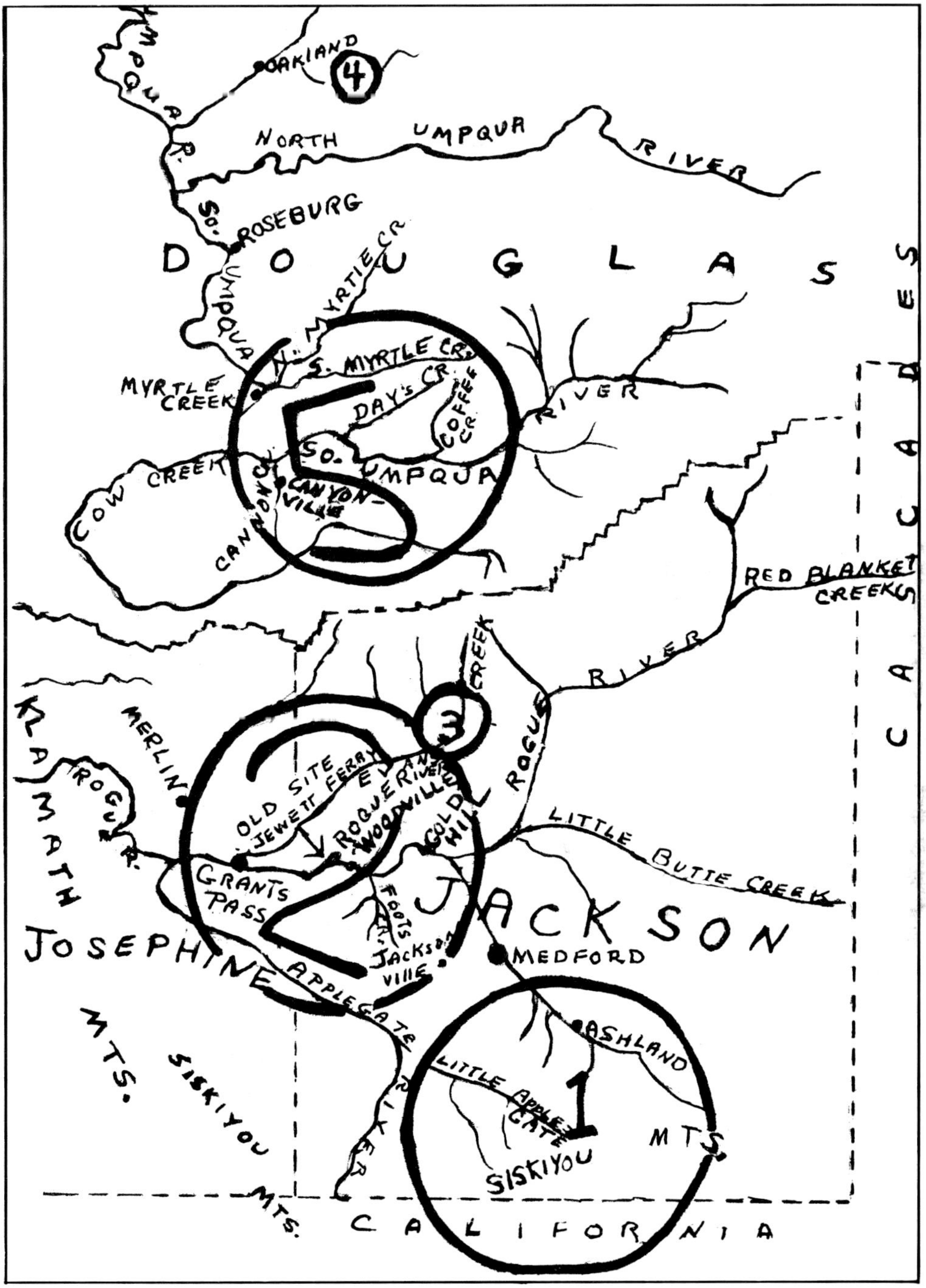

LOST MINES OF SOUTHWEST OREGON

1—*Lost Vault of Gold*—General area of search
2—*Indian Lost Cinnabar Mine*—General area of search
3—Chisholm group mercury mines
4—Nonpareil Mine
5—*Ed Schieffelin's Lost Oregon Gold Mine*—General area of search

Left: The poet, Samuel L. Simpson, "Oregon's Sweetest Singer," hunted the Lost Vault of Gold and later wrote the story of that search and its tragic consequences. *Right:* Ed Schieffelin, one of the most romantic figures in western mining, was the discoverer of the famed Tombstone, Arizona, silver mines, and the loser, through death, of a rich Oregon gold mine. (Jacksonville, Oregon, Museum)

"Papoose" rocker and tin-can dipper—a portable, one-man gold-mining outfit, copied by Al Vanderpoel from one used by an old prospector. "Fifty times more efficient than the gold pan," says Al.

Solid silver container dug from a shipwreck site off the Nehalem Spit in 1898 by the John Tuttle family of Tillamook, Oregon, and now owned by Mr. and Mrs. Harry Tuttle of Winthrop, Washington. On lid is hallmark used by European silversmiths around 1750. (Burford Wilkerson)

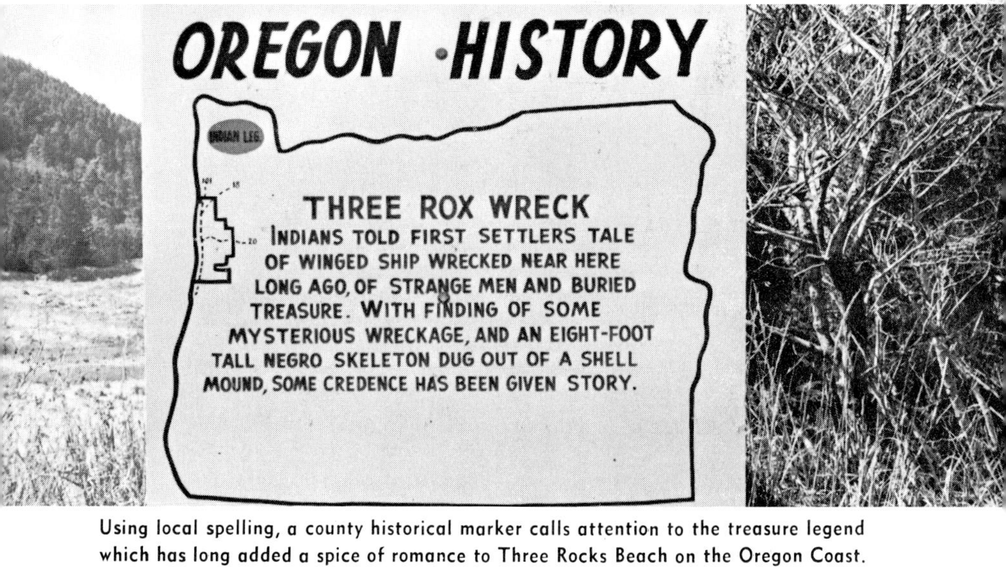

Using local spelling, a county historical marker calls attention to the treasure legend which has long added a spice of romance to Three Rocks Beach on the Oregon Coast.

Left: With Neahkahnie Mountain looming in the background, Alex Walker, curator of the Tillamook County Pioneer Museum, holds the famed block of beeswax from which a piece was recently taken for the purpose of radiocarbon dating. (Burford Wilkerson) *Right:* Weather-beaten ship's knee, here displayed by Ray Lindsay of Garibaldi, Oregon, was found where the shifting mouth of the Nehalem River once ran and where mysteries of beeswax and buried treasure chests still abound.

Tony Mareno, shown operating a caterpillar tractor during his controversial dig for the Neahkahnie Treasure, was constantly surrounded by curious onlookers at Manzanita Beach on the Oregon coast.

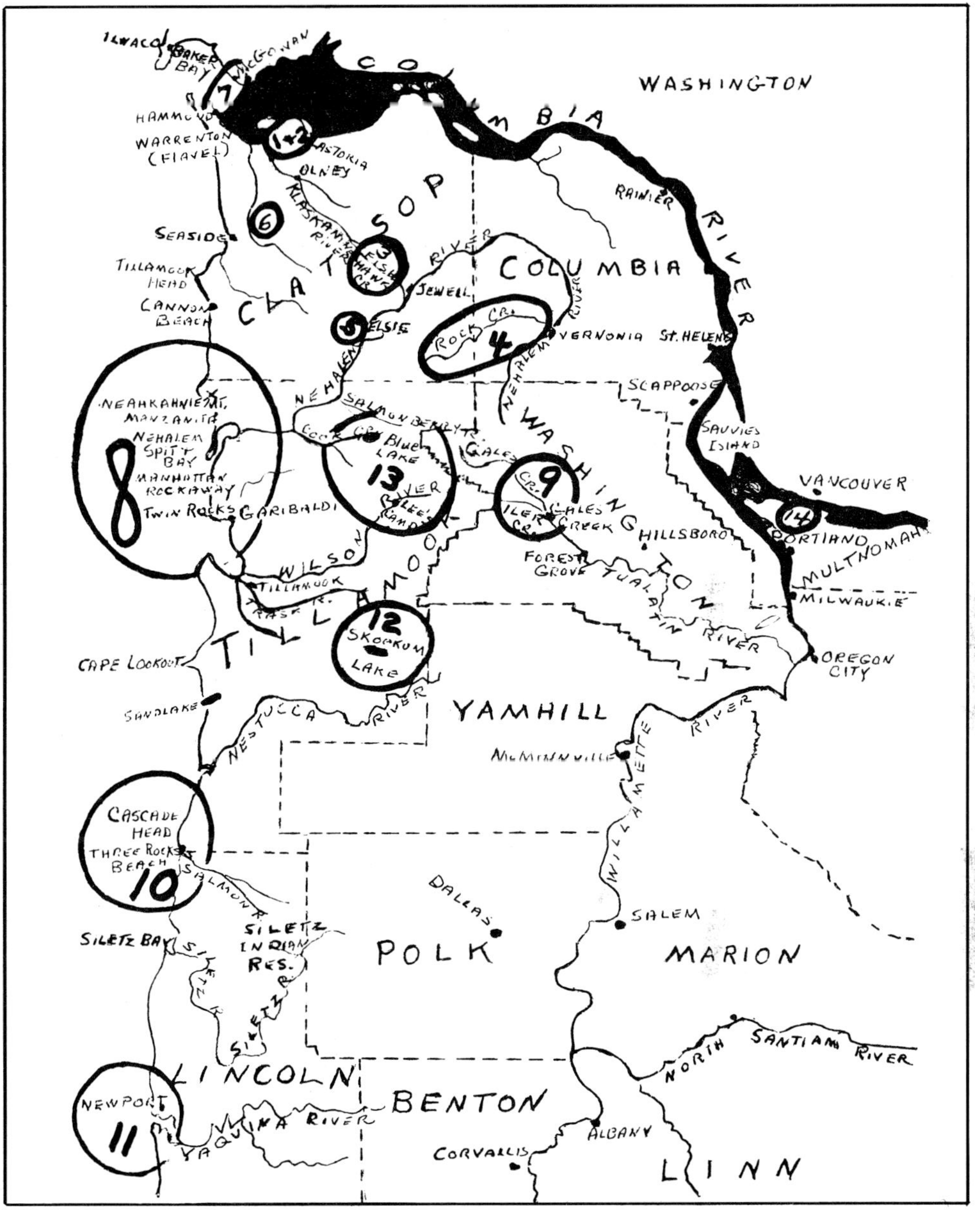

NORTHWEST OREGON

1 & 2—Coxcomb Hill Cache and Flavel House Cache, Astoria
3—Fishhawk Falls Lost Mine
4—Rock Creek Mining District
5—Cruiser's Gulch, site of old gold strike
6—Clatsop Plain Mine
7—Magnetite deposits
8—Many mysteries of Neahkahnie and Nehalem
9—Iler Creek, where Portuguese sailor found refuge
10—Pirate chest of Three Rocks Beach
11—Place of discovery of old anchor chain and other shipwreck evidence
12 & 13—Possible locations of Lost Tillamook Mine
14—Portland Treasure Chart

Don Viles of Garibaldi, Oregon, prospecting a stream in the Oregon Coast Range while seeking the Lost Tillamook Mine.

Rugged country of the Tillamook Burn in the Oregon Coast Range, land of the Lost Tillamook Mine.

Cook Creek, a lovely entrance to the Tillamook Burn in the Oregon Coast Range, where many seek the elusive Lost Tillamook Mine.

The great Tillamook Burn of 1933 wrought tremendous changes in the land of the Lost Tillamook Mine.

Long-sought guide to the true location of the Blue Bucket Mine is the gravestone of Sarah Chambers, discovered in 1877 in the Malheur country. Westward from her gravesite one day's travel, or two, the Joe Meek wagon train in 1845 camped at the creek where the gold was picked up.

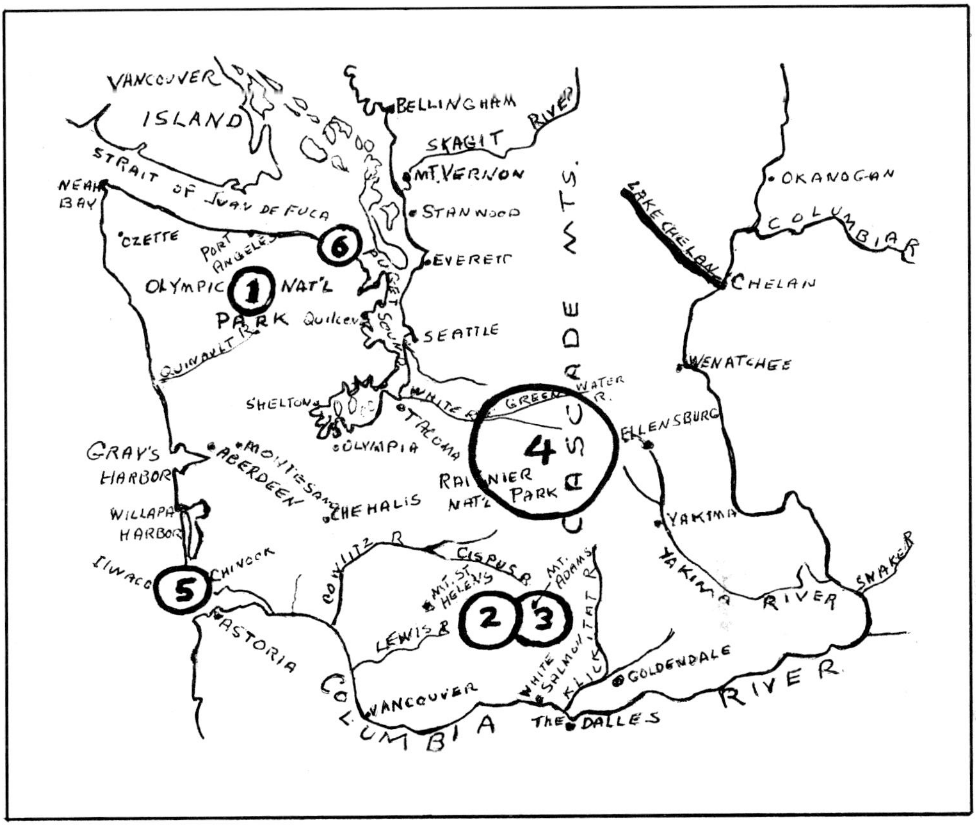

WESTERN AND CENTRAL WASHINGTON

1—Lost Gold Strikes of the Olympic Mountains

2—Rabado's Lost Gold Mine

3—Cave of the Hidden Mine

4—Chief Kitsap's Gold Mine

5—Baker's Bay, the Captains' Caches

6—Chevy Chase Treasure Chest

Tull City copper mining development, on eastern slope of the Olympic Mountains, came at the same time as rumors of lost gold prospects.

Burro train of dynamite ascending Dirty Face Ridge in the Olympic Mountains, on way to Tull City mines.

But illness woke the red-daubed braves long before dawn. The peculiar properties of mercury which make it a corrosive poison upon the system of man and even of animals was working its horrors. They woke up shaking and shivering, their mouths and throats inflamed, their intestines griped and painful. Nauseated, diarrheic, some of the warriors felt near death. Even their horses were found shaking and sick.

One of the few adult males not affected was a captive Klamath slave, he of course not having been allowed to participate in the painting rite. Seeing the Rogue camp disorganized by the sudden illness which had struck it he kept his wits about him and was able to slip away unnoticed.

Too physically disabled to take the warpath, the best the Rogues could do was to gather themselves and their animals together for the painful journey back to their own illahee by the river. In contrast to their eager march of the day before they dragged themselves homeward for two days. Even after reaching their own campground, the males still groaned and writhed all the first night. And at dawn the Klamaths fell upon them. For the escaped slave had sped back to his people with the news of the incapacitated state of the usually powerful Rogues. A Klamath war party was hastily assembled and led to the scene by the slave.

With the Rogue warriors still ill and taken by surprise the carnage was terrible. Nearly half the tribe was killed and it would have been entirely wiped out had not a raiding party of upper Rogues happened by to engage the Klamaths in battle and put them to flight.

This was how he knew the red rock was an evil thing, Indian John said in finishing his story. It was bad medicine, bringing sickness and death, and he wanted to talk of it no more.

Robert Jewett tried to make him see that his tribesmen had daubed on too much of the mercury too carelessly, but it was no use. John had explained to his white friend why he could not reveal the location of the ledge of red rock and he would say no more.

So Jewett was forced to report his failure to Mr. D. and his other anxious copartner and though they now had a tidy sum invested in the Indian there was nothing to do but take him back to the reservation.

But this by no means put an end to the Jewetts' dreams of wealth. With Douthitt they moved a quartz mill to the Elizabeth Mine and took out $40,000 in gold. Even after they left the ferry in 1864 they continued their interest in finding the Indian Lost Cinnabar Mine.

Right after quitting the Rogue River, Robert Jewett ran a hotel in Portland. He then bought a ferry on the Burnt River in eastern Oregon, residing there for twelve years. So either Thomas Jewett continued to prospect or the two brothers grubstaked others, for in 1875 they made a rich cinnabar strike in Douglas County, ten or twelve miles east of Oakland. They named the mine the Nonpareil, meaning "none better," and erected a furnace capable of retorting 600 pounds of ore a day. Returns were so promising they were able to sell the mine to the New Idria Company at a good profit to themselves.

In discovering the Nonpareil had the Jewetts located the Indian Lost Cinnabar Mine? Was it the happy ending to their long search? H. H. Bancroft in his *History of Oregon* states that in making the Douglas County strike the Jewetts unravelled the mystery of Indian John's amazing specimen. But this simply cannot be true.

At the time the Nonpareil discovery was stirring up renewed interest in mercury mines the man we have called Mr. D. wrote an account entitled, "Cinnabar in Southern Oregon—The Unfound Ledge," which was published in the September 1875 issue of the magazine *West Shore*. It tells the story of Robert Jewett's adventures—or misadventures—in trying to pry out of Indian John the secret of the rich specimen. It begins:

"For months past, word has been coming from portions of Southern Oregon of the occasional finding of ledges or mines of cinnabar—the red sulphur of mercury, or quicksilver. As yet these discoveries have been made mostly in Douglas county and in the southern portion of it. Only one of two 'finds' have been reported in Jackson county, and neither of them of much account, apparently. But there is in that country somewhere a vein or mine, not yet struck, from which was taken, years ago, the richest piece of cinnabar ever found on the whole Pacific coast . . . not excepting any from the famous mines at New Almaden . . . or New Idria. Just where this ledge or mine is, however, is not solved. I am sure of this, though, that it is not above twenty-five miles from what used to be known as Jewett's ferry, on Rogue River . . . ."

Thus both the title of the article and the text prove that the Indian Lost Cinnabar Mine was still regarded as lost.

Twenty-five miles would seem to be a good estimate of the distance described as one day's chase or march away from the main Indian encampment on the Rogue River. The location of the Nonpareil Mine in the Umpqua River drainage is three times that distance. So it could not possibly be the ledge from which John's specimen came.

Three years after the discovery of the Nonpareil, in 1878, a quite sensational cinnabar strike was made on upper Evans Creek about fifteen miles northeast of the old site of the Jewett ferry. The find was made in a meadow on the south slope of the Umpqua Mountains in a vein or dyke which experts have described as extending clear across that range into that part of Douglas County demonstrated as having some good cinnabar deposits.

The mercury mines staked on Evans Creek became known as the Chisholm Group. Were the Chisholm Mines, then, really the Indian Lost Cinnabar Mine? All we can say is that while the Chisholm Group were fine mines their ore never did retort mercury in as great quantities as the New Almaden. Nor has any other mercury mine discovered in Jackson county produced quicksilver in quantities sufficient to make it a candidate for the lost mine. So presumably the richest cinnabar mine in the United States and perhaps the world still awaits discovery.

A circle with a radius of twenty-five miles, with center near the mouth of Evans Creek on the Rogue River, would include an impossibly large amount of territory. A direction of search is plainly needed. Perhaps that direction is indicated if the treasure hunter remembers that extensive vein or dyke of cinnabar ore beginning in the upper reaches of Evans Creek and extending north over the summit of the Umpqua Mountains. From some outcropping in that area may have come the Indian gift to Jewett of the red rock specimen with mercury content so fantastically high as to outshine anything ever seen even at the New Almaden or the New Idria.

If that ledge could be discovered, Jacksonville would bloom again, with the shades of Robert Jewett and his co-conspirators nodding approvingly and the shade of Indian John stirring uneasily with memories of race anxiety.

Southwest Oregon

ED SCHIEFFELIN'S LOST OREGON GOLD MINE

Red Blanket Mine; Coffee Creek Mine

If ever there was an old time prospector always over the hill in pursuit of the will-o'-the-wisp and never allowing himself to be diverted from his drive toward fortune it was Edward Schieffelin.

Probably his obsession began in 1852 when he was only five years old. It was then that his father, Clinton Schieffelin, and an uncle, Joe Walker, left the family home in Tioga county, Pennsylvania, for the gold fields of California: surely an event to stir the mind of an active youngster.

By the time the men reached California the words on everyone's lips were "Rogue River" and "Jacksonville." So they continued on to Oregon and in the fall of 1853 bought a land claim on Rogue River near the Jewett ferry and made their living farming and mining. Clinton Schieffelin also filed on a donation claim close to the ferry and sent for his wife and children, who joined him on Rogue River in February 1857.

Ed was not yet ten years old when this western jump was made and he was all agog over the opportunities the new place presented. "On arriving in Oregon," he says in a memoir, "the first thing I did was to get a shovel and milk pan and go down to the bank of the river looking for gold." He was thrilled to discover glittering flakes in the water and he worked hard to accumulate a teaspoon of them, only to have his Uncle Joe laugh amusedly and tell him the stuff was only mica or fool's gold.

This by no means discouraged Ed. He was only determined to find the real thing. At ages eleven, twelve and thirteen he frequently took pick, shovel and pan, going "off into the hills in the neighborhood, washing gold in the gulches," and whenever he went deer hunting he stopped constantly to "break the quartz and look for gold in it."

He was only fourteen years old when the news of the Salmon River strike in Idaho reached the Rogue and he became so excited he ran away to join the stampede, only to be apprehended sixty miles away by a neighbor and brought back home. At age seventeen he was determined to leave for the Montana gold fields but found his parents so opposed to the plan he gave in and consented to begin his mining career closer to home.

For the Schieffelins as for their friends and neighbors, the Jewetts, the center of all things was Jacksonville — and Jacksonville managed to keep things lively. Strikes—rich or not so rich —were made on Williams Creek, Galice Creek, Foots Creek, and others over the divide in the South Umpqua country. Hopes still ran high and there was hardly a creek without some miners or prospectors trying it out.

So Ed was able to find work at a placer claim only five miles away. After a month's labor he drew his wages and bought a part interest in another claim. He worked the property for three months, until the season's water gave out. He drove a freight wagon until the water came in but found the claim simply wouldn't pay.

He went to work for wages but as soon as he had built up a small stake, away he went prospecting again. He found one place that looked good "and as it was in rather heavy ground and required considerable work to get to it, I built a cabin and got some tools, provisions, etc., ready for another trial. But after working awhile . . . . I concluded it would not pay . . . ."

Ed does not say just where these particular episodes took place but it seems sure that between his seventeenth and twenty-first years he ranged all the mining fields of southwest Oregon, from the coast clear up into the Umpqua drainage. Nor was he the only one in his family interested in mining. His father bought and sold a share in the Jewett lode and later owned ditches and parts of claims on Foots Creek. Two or more of his brothers later became miners.

Around them others were making their fortunes while luck kept passing the Schieffelins by. In 1869 Ed was fed up with Oregon and his parents could no longer restrain him from striking out into the world.

He began a roving, prospecting life which took him to the Surprise Valley, Nevada, and on to the Owens River, California; through the Pioche country in southern Nevada, to the Salt Lake district and Corrine, in Utah; up to the Snake River and into the Boise Basin. Not liking Idaho he returned to winter in Nevada, then started south into the Grand Canyon of the Colorado. He made yet another trip north and south, wintering this time in Texas.

Always he was painfully broke. In Utah his funds ran out "and I sold my saddle mule and put my blankets on my back . . . ." In Arizona he was so poor another man had to pay his twenty-five cent fare across a Colorado River ferry. In Texas he and his brother chopped cordwood for eating money. In his reminiscences he says bitterly that at a time when wages were high and other people were all making money he could seem to make none. He became ill and returned home to Oregon, arriving with $2.50 in his pocket.

During his convalescence he did some serious stocktaking. It was probably the first time in his restless, questing life that he had sat still and thought hard. He believed he saw where his trouble lay. He had listened to too many passing rumors, followed too many hair-brained mining excitements, paid too much attention to the advice of others and not enough to his own intuitions. When he hit the trail again he vowed to act differently. He would follow his own instincts, lay out a course and stick to it—never turn aside until the prize was won.

At the end of three weeks he borrowed $100 from his father and set out for Arizona. He got only as far as southeastern California—broke with his blankets on his back. Spurred by his new sense of purpose he found a job and worked diligently but it took him fourteen long months to save enough to buy the necessary prospecting outfit: two good mules, saddles, guns and tools. "I got these together with the determination of never parting with them again," he says. "I had a good outfit, and made up my mind, no matter what occurred, I would keep it, and if I could not do any better I would live off my rifle until I found a prospect."

With that hard-headed resolution he set off into Arizona in January 1877 and within six months he had struck it rich, discovering the fabulous silver ledges of Tombstone. With that strike he not only became a wealthy man but one of the most important, most romantic figures in all western mining lore.

Not that it had come easy. First he had gone to the Grand Canyon country, prospecting and finding nothing. He then decided to trail along with some Hualapai Indians who had been recruited for Apache scouting in southern Arizona. Schieffelin's own account proves conclusively that he was not a scout as so many writers have stated; he simply joined the party for protection through the Indian country. At Camp Huachuca, commanded by the noted Al Sieber, Ed found traveling out with army parties so slow and cumbersome he began making prospecting trips alone. This was dangerous practice in Apache country where murder of travelers and settlers was common but Ed ignored warnings and went right on prospecting, saying he would surely find something in the country. The soldiers at the camp said, "Yes, you'll find your tombstone."

So on discovering his first silver prospect he named the claim "The Tombstone." But again he was desperately broke and had to go off to Globe, Arizona, to hunt up his brother Al to get help in developing the lead. Fortunately he also was able to interest another man, Richard Gird, an engineer and assayer, in his find. The three returned to the ledge and located the Lucky Cuss, the Tough Nut, and another claim whose title later being disputed became known as the Contention.

Schieffelin's own account disproves two hoary old tales about his great strike: He did not come across any skeletons with piles of silver ore beside them. He did not follow a rabbit up the slope to where the silver ledge was found. It was a piece of float which attracted his attention and which he traced up the gulch.

In March 1880 the Schieffelin brothers sold out their interests in the Tombstone claims for the low figure of $600,000. But Richard Gird retained his interest and it appears that when the silver mines began producing their tremendous wealth he cut the Schieffelin brothers in on further profits. Ed Schieffelin was only thirty-two years old and his lean days were over. During the rest of his life he never lacked for money. He bought a big house in Los Angeles and brought his parents down from Oregon. He traveled to New York and learned to enjoy the comfort of luxury hotels.

But a man whose life has been one long quest cannot immediately settle down to leisure and in 1881 Ed was in Portland, Oregon, spending large amounts of money in building a sternwheel steamer for a prospecting trip to Alaska. In this vessel and a chartered schooner he and four other men prospected the Yukon River during the summers of 1882 and 1883. But the Klondike strike was not to be theirs. They disbanded without finding any evidences of gold.

In the meantime Ed had become enamored of Mary E. Brown, a native of Virginia who had been living in San Francisco. In the fall of 1883 Mrs. Brown traveled to Albuquerque, New Mexico, to meet Ed and they were married at LaJunta, Colorado. The next year the couple bought and settled into a large house at 1511 Central Avenue in Alameda, California.

In about 1887 a representative of the H.H. Bancroft Company came to call upon Schieffelin and set down his reminiscences concerning the discovery of the Tombstone silver mines. The scribe was much struck by Ed, describing him as large, powerful, bronzed, with keen blue eyes and brown beard and hair—"a type of the physically perfect man." And he found the miner who had known so many years of grinding poverty now "surrounded by the comforts and elegances of a delightful home."

Yet marriage and domesticity could not settle Schieffelin down entirely. He admitted to having made two or three recent prospecting trips to various localities—and he continued to be a rover.

The review of his life set him to thinking of Oregon and his first amateurish mining attempts. In speaking of that one early claim where he had built himself a cabin and worked several months before he gave up he said, "I have since thought I made a mistake and that there was gold there but in the bedrock instead of in the gravel; if I had worked up the bedrock I might have got something out of it."

He developed an itch to go back and apply his hard-earned, expert knowledge to places and prospects he had passed over too quickly. So either late in the 1880s or early 1890s he turned his attention back to Oregon, the land of his youth and young manhood.

Most authors give the time of Schieffelin's return to Oregon as the middle or late 1890s (1895, 1896 or 1897 are usually mentioned) but this information is incorrect. In a mining transaction recorded in Jacksonville on October 9, 1893, Ed is referred to as "of Jackson County," and he may have returned to the state long before this piece of business was recorded. Possibly he began his resurvey of Oregon mining fields soon after old memories were stirred up by his sessions with the Bancroft historian.

During his years of absence a little town had grown up at the mouth of Evans Creek near where the old Jewett ferry had crossed the Rogue. It was first called derisively Tailholt but later took the more dignified name of Woodville. In 1913 it became Rogue River, the name which appears on maps today.

The visit of so famous a man back to the locality where he and his family were well known was certainly an event of note. Yet there seems little but fantasy in the picture drawn in bold strokes by one treasure writer who describes Schieffelin as dashing dramatically down Woodville's main street in a bright blue Concord coach with yellow-gold running gear, booming "Whoa!" to his four-horse team of matched sorrels and leaping down to demand, "Where can I get some good whiskey?"

That Ed Schieffelin ever traveled in such circus style is impossible to believe. He knew his worth and was not unduly modest but always he was a man of taste and discretion who did not call unnecessary attention to himself. And he seems not to have been a drinking man at all.

But it is apparently true that in Woodville, though perhaps not at the time of his first return, Ed hired husky young Charlie

Warren to go out with him on prospecting trips to drive team and act as camptender.

And of course Ed revisited Jacksonville both for purposes of business and pleasure. In the intervening years the mining town had known both triumph and tragedy. Its most notable piece of success, still much boasted of, had taken place in September 1880 when it had played host to the President of the United States, Rutherford B. Hayes, who with his party had stayed overnight in the new, swank United States Hotel. Its most heartrending failure was in being bypassed by the railroad and seeing upstart Medford, only a few miles away, growing up as rail center. But J'ville had built its own railroad branchline completed in 1891 and dubbed the "Cannonball Express." So the town was still full of beans and dreams, and still county seat.

The studio of Peter Britt, with modest beginnings in gold rush times, was now one of Jacksonville's solidest places of business with a wide reputation for fine photography. To it Ed Schieffelin repaired to have his likeness taken, not in the guise of the rich man he had become but of the prospector he remained at heart. For the occasion he dressed in his miner's clothes and brought with him his miner's pick, his old canteen and his rifle.

From the Peter Britt photos, one of which appears in this book, we know how Ed Schieffelin looked in the mid-nineties but unfortunately we cannot trace in detail or with assurance his movements during this period of prospecting in Oregon. For in contrast to the first forty years of his life, so carefully sketched by the Tombstone reminiscence, the last decade is relatively barren of facts. We do have some little evidence from mining records, we have a bit more from his last will and testament, we have a surviving story or two which ring true, but mostly we have a snarl of lost mine tales, many of them in seeming contradiction to each other.

According to Charlie Warren, Ed's teamster and camptender, his boss's restlessness took them on large circling prospecting trips, all the way from Coquille on the Oregon coast, down into Nevada and California, and back again into Oregon as far north as Roseburg. Sometimes they spent weeks in the hills. Said Charlie:

"At night, after going to bed, Ed would think out what he would do on the morrow. Awakening at daylight his mind was set like a steel trap, and he would follow the plan of the night before. It was this bull-headed trait that brought him success where other men had failed."* Which is certainly a true-to-life description of the way Ed's mind worked after his vow to banish all impulse.

Several times in their wanderings they found ground with enough indications of gold to impress the ordinary man but Ed only shrugged and passed on. Unless a strike was as big as his Tombstone discovery he didn't care to be bothered. He was already rich. What he wanted now was the glory of another bonanza to his credit. Or so Charlie Warren said.

Actually old mining records reveal that Ed invested money in what appears to be quite ordinary mining property. In the October 1893 transaction he bought two claims, the Delusion and the Phantom Quartz, from his brother Charles L. Schieffelin, the only member of his family still living in the Woodville area. The claims, which had been located and filed on by Charles the previous January, were described as being "on the east side 500 feet from the left hand fork of Foots Creek, one half mile above George Lance's cabin, and near the trail to the middle fork of Foots Creek." The price Ed paid was $1200.

Foots Creek is a southern tributary which enters the Rogue River a little east of the town of Rogue River (old Woodville). It was the scene of early mining activity. Ed's father at one time owned some ditch and other rights there and Ed himself had doubtless prospected the area.

And fifteen months later Ed Schieffelin still displayed an interest in the Foots Creek Mining District. Records show that in January 1895 he bought an undivided 3/5th interest in two quartz lodes and a quartz mill site for $10 from Alex and Jane Orme, who appear to have been friends of his.

But mining records are at best bare bones and it is possible that Ed visualized big returns from these investments. More probable it is that he was helping out his brother in the one

———————

* From the article "Lost: Ed's Red Blanket and Rich Gold Mine," by Dale Vincent, Portland *Oregonian*, March 31, 1946, magazine section, page 8. Quoted by permission of Viola Vincent.

instance and lending the weight of his name to the Ormes in the other.

On September 16, 1896 he was back home in Alameda. On that day he drew up and signed his last will and testament, leaving half of his property to his wife and half to his brother Jay and making his brother Charles and his wife co-executors of his estate. Then either late in 1896 or early in 1897 he headed back north.

In the spring of 1897 he chose to prospect the tributaries of the South Umpqua in Douglas County. Finding a cabin near the head of Day's Creek which was still habitable he decided to make it his headquarters while examining the surrounding country. He made arrangements to leave his sorrels and coach with a rancher near the mouth of the creek. Then he and Charlie Warren, who was still working for him, packed everything needed up to the cabin: Ed's camping outfit, his prospecting tools and enough supplies to last for a couple of weeks. As soon as he was settled in he allowed Charlie to go on vacation to pay a visit to his family. Ed was left alone at the cabin.

The rancher had been told the approximate date Ed would be hiking down to pick up more supplies. When the prospector failed to appear on schedule the man began to worry. As it happened, Alex Orme came along, wanting to talk to Ed. On hearing he was overdue Alex became alarmed and immediately hiked up Day's Creek to the cabin. There he found Ed Schieffelin dead.

The news was published in the Ashland, Oregon, *Tidings* on May 17, 1897, headlined "ED SCHIEFFELIN'S LONELY DEATH":

"Edward Schieffelin, the well-known prospector and one of the original discoverers of the famous Tombstone Mine in Arizona a number of years ago, was found dead at his cabin about 20 miles east of Canyonville last Friday. Justice Stock immediately proceeded to the place and held an inquest, the verdict of which was that he came to his death from natural causes, presumably heart failure, as no marks of violence appeared on his body. He had on his person $68.75 in coin and a splendid gold watch. He appears to have been dead since last

Tuesday. The body was in bad condition and he was buried by the jury.

"Ed Schieffelin was raised on a farm on Rogue River in Jackson County, and his career is one full of romance and adventure rarely experienced. The *Tidings* will devote more space to it in another issue."

For some reason the promised sequel never appeared and these are all the hard facts we have to go on in regard to Schieffelin's death. All the rest is legend.

As the tale is told Ed was found face down beside a mortar and pestle where he had been grinding ore. A jar standing nearby contained samples of the same ore which later assayed $2000 a ton in gold. And in a diary which Schieffelin had been keeping the last entry read, "Struck her rich again, by God!"

Charlie Warren is credited with a story to the effect that Ed had with him two good wool blankets, one red and one blue, and whenever his prospecting trips took him far from the cabin he had packed a blanket along so he could sleep out overnight. On his death the red blanket was nowhere to be found. From this Charlie deduced that the rich ore had been found far enough away that Ed had stayed overnight when he had visited the diggings and had left the red blanket there expecting to return. So the thing to do was look for Ed's last camp site. Whoever could find the red blanket would find that fabulous last prospect—hence the name Red Blanket Mine.

In another version of this same tale it is not a blanket but some cooking pots which were missing from the cabin, indicating that Ed had camped at his diggings. Still, these two statements are not necessarily in conflict since if Ed really did establish an overnight camp on the scene of his strike he might well have left both his blanket and some cooking pots there.

In a quite different story, often repeated, there is no mention of diary entry, red blanket or cooking pots. In this story the clues to Ed's last strike are a letter mailed two weeks before his death, and also a treasure map found in the cabin. In the letter he wrote, "I have found stuff here in Oregon that will make Tombstone look like salt. THIS IS GOLD." The map gave directions for finding the new golden bonanza.

As told by the popular writer, Stewart H. Holbrook, this map passed to a nephew of Schieffelin's. While serving in the Canadian forces during World War I, this nephew was wounded at Verdun and gave the map to a buddy before he died. Holbrook writes:

"I know this man and have seen the paper . . . . The writing is beyond doubt that of old Ed Schieffelin. Two places on the map are marked 'Here.' The problem presented by the map is to know certain distances that are outlined. These are in cypher of Schieffelin's own making and can be learned, apparently, only by experiment—which takes a good deal of time in a region as large and as rough as Douglas County."

Of course there may be truth in *all* the different versions. For if Schieffelin was excited about his big new strike he may well have written a letter, drew a map and noted the event in his diary. Blankets and cooking pots may have been left at the diggings.

Efforts to locate the diary and the map have met with no success. They may still exist, however.

The most disappointing thing is that no very specific clues have been left as to the location of the cabin where Schieffelin died—only such vague designations as "20 miles east of Canyonville" and "at the head of Day's Creek." Since Day's Creek is about fifteen miles long with a fairly large headwaters area, that isn't much help.

R.W. McAllister in his *Lost Mines of California and the Southwest* refers to Schieffelin's lost Oregon prospect as the "Coffee Creek Mine," and says Ed went to Douglas County "to look over Coffee Creek where he had panned gold as a kid." This makes a good deal of sense. The Coffee Creek gold fields were much talked of following their discovery in 1858 and it would have been natural for Schieffelin to have prospected the area in his youth.

Both Day's Creek and Coffee Creek are northern tributaries to the South Umpqua, Coffee Creek being the next stream east of Day's. Their headwaters closely adjoin but due to a western swing made by Day's Creek in its course their mouths are a dozen miles apart. Still, the distances are such that from the Day's Creek cabin where Schieffelin spent his last weeks he could easily have prospected Coffee Creek, and McAllister may be right: the lost mine could lie somewhere along Coffee Creek.

Confusingly enough, the map which accompanies Dale Vincent's *Oregonian* article, "Lost: Ed's Red Blanket and Rich Gold Mine," has an arrow pointing to a shaded area along Canyon Creek, with a notation reading, "Canyon Creek Country—Where Schieffelin made his last gold strike." Canyon Creek flows into the South Umpqua from the south. From its mouth to the head of Day's Creek is a good twenty miles as the crow flies and much farther for a man legging it over rough country. It is inconceivable that Schieffelin would have attempted to prospect Canyon Creek while living at the head of Day's Creek. The text of Dale Vincent's article does not put the lost mine in the Canyon Creek drainage yet the author knew the South Umpqua country well and his map may illustrate a theory he had developed.

Another odd thing: A tributary to the middle fork of the Rogue River is named the Red Blanket Creek. The location is forty or fifty miles from Day's Creek, seemingly too far away to have any connection with Ed Schieffelin's missing red blanket. How did the name happen to be placed there?

There is some evidence that Ed's brother located or believed he located the place from which those last ore samples came. For mining records show that on September 15, 1900 Charles L. Schieffelin filed on a quartz claim, together with a north extension and a south extension, said to bear gold, silver or lead. This property was located on the east side of Day's Creek (though no more exact designation is given) and was named the Last Prospect.

Last Prospect—surely that reveals Charles' conviction that these were his brother's last diggings. Yet no other word is heard of these Day's Creek claims and obviously they failed to live up to their promise. So the Schieffelin gold mine is still regarded as lost. For more than seventy years men have continued to tramp the wilds of Douglas County in the hopes of reechoing that last immortal line, "Struck her rich again, by God!"

When Ed Schieffelin's will was read it was found to contain one unusual paragraph:

"It is my wish, if convenient, to be buried in the garb of a prospector, my old pick and canteen with me, on top of the granite hills about three miles westerly from the city of Tombstone, Ariz., and a monument such as prospectors build when locating a mining claim, built over my grave, and no other monument or slab erected. And that none of my friends wear crepe. Under no circumstances do I want to be buried in any cemetery or graveyard."

The body of Schieffelin was disinterred and shipped to Tombstone, where it was reburied in compliance with the miner's request. The prospector's monument raised over his grave is still there, three miles west of that tough old mining center which is now a virtual ghost town.

As was true of Tombstone, Jacksonville lived to see its days of glory depart. It was Medford on the main rail line which grew and flourished while Jacksonville stood still and then began to slide backwards. In 1927 when aggressive Medford stole away the county seat Jacksonville was finished, or should have been and many thought it out for the count. Certainly commerce had told it an abrupt goodbye.

Yet there remained an invaluable resource no one could take from it, a resource its citizens proved happily to appreciate: its pivotal role in the history of southern Oregon. Wisely Jacksonville was turned into an outdoor museum.

Along its tree-shaded streets one can see preserved buildings so old as to represent the original discovery; the building used as a refuge in times of Indian trouble; the old Table Rock Saloon; the gingerbreaded glory of mansions built with gold rush money; the utilitarian plainness, now so quaint, of the Beekman Bank (opened by "Beek" the expressman); the fancy United States Hotel where a president slept. In all a half-hundred buildings make up this shrine to the mining past.

Within this setting the formal Jacksonville Museum in the old brick-and-stone court house is but a part of a larger whole, and within that a still smaller but equally important part: a replica of the photo gallery of Peter Britt, showing his early cameras, trappings and equipment, together with hundreds of his priceless collection of pioneer photographs. That Ed Schieffelin is represented we know. But the collection being so vast as to have not yet been completely catalogued we cannot say for sure, but because almost everyone sat to Britt the likenesses of the Jewett brothers and their friends, Sam Simpson and Ted Harper, and the earlier Wilson brothers may all turn up yet. It does not matter. In history they are forever part of that magnificent horde of treasure hunting romantics and prospectors whose rich gulches, whether real or misty as dreams, created memorable old J'ville—and Oregon.

Northwest Oregon

COXCOMB HILL CACHE

Astoria, oldest settlement in old Oregon, located just inside the Columbia River mouth, is famed for its proximity to Fort Clatsop, the winter home of the Lewis and Clark Expedition in 1805-1806; famed as the site of the first American fur trading post on the Pacific Coast, built in 1811 for John Jacob Astor; and steeped in storied romance for generations of readers of the writings of Washington Irving.

These were auspicious beginnings, unfortunately soon lost in the helter-skelter of pioneer development. As the early American town grew, "buildings were thrown up without much thought of durability; so eager were the early residents to secure river frontage that much of the town was constructed on planked-over pilings extending into the river" — according to Emma Gene Miller in her excellent book, *Clatsop County*.

In the early 1870s Charles Nordhoff, a famous writer of the day, visited Astoria in the expectation of finding it enveloped in the glamorous aura of its historic past, but his disappointment was so extreme he wished he had passed it by unseen. "It looks like the rudest Western clearing you ever saw," he wrote. "Its brief streets are paved with wood; its inhabitants wear their trousers in their boots; if you step off the pavement you go deep in the mud; and ten minutes' walk brings you to the forest primeval . . . ." — and there was so much forest primeval that he echoed the sentiments of a man who was heard to exclaim, "Timber!—timber till you can't sleep!"

And Helen Hunt Jackson, author of *Ramona*, who called in at Astoria in 1881, wrote that "it must be classed as amphibious, being literally half on land and half on water . . . . Long wooden wharves, more streets than wharves, fairly fringe the shore; and the street nearest the water is little more than a succession of bridges from wharf to wharf. Frequent bays and inlets make up, leaving unsightly muddy wastes when the tide goes out" — and it amused her to see family washings hung out on lines over the tide flats and children playing along the muddy margins.

Today Astoria is a charming little city which has long since outgrown its undignified past. Though not exactly aquatic it is surrounded on two, and almost three, sides by the waters of the Columbia River and Young's Bay. It has mills, a fishing fleet, canneries, and a new bridge leading to the north shore and Washington State.

The most eye-catching structure in town is one which strikes the visitor at first glance as unbelievable. For the last thing anyone expects to see here at western track's end is a Roman column looming on top of a hill.

Patterned after the noted Trajan column of Rome, erected in 114 A.D. by the Emperor Trajan, the tower rises 125 feet above Coxcomb Hill, a mesa-like eminence which forms a backdrop to the Astoria business district and is itself a residential area. Inside the tower is a circular stairway of 166 steps leading to a topside platfom which gives a tremendous view of the whole world of the river mouth.

This amazing structure is known as the Astoria Column and was erected in 1926 with money supplied by the Great Northern Railway and the descendants of John Jacob Astor. Decorating the outside of the upthrust shaft is a spiral frieze depicting the history of the area. Each scene is emblazoned in brilliant colors: the forest primeval which so oppressed Nordhoff; the discovery of the Columbia River by Captain Robert Gray in 1792; the arrival of the Lewis and Clark Expedition on the Pacific shore; the building of Fort Astoria; the Fort's passing to the British in 1813 and being restored to the United States in 1818; and the coming of civilization.

What is *not shown* is a local piece of history which has stirred more modern-day Astorians to excitement than all those pictured events put together: the story of the Coxcomb Hill Cache. And interestingly enough, the focal point in the treasure tale is along the very edge of the hill on which the column stands.

So, when you visit Astoria and go see the beautiful Trajan column (and you must see it, as it is unlike anything else in America), you will find yourself driving up Coxcomb Hill Road past the site of the cache.

As you round the second curve of the road pause a moment (if there is no one behind you and there is room to pull up in safety) and try to envision the hillside before the present road was put through.

Back beyond the bank to your left, fifty feet or so into the brush, stood a cabin where a nondescript old man lived in the 1920s. In those days he was monarch of all he surveyed as he looked across the ungouged terrain to the right of the present road into a steep ravine, no longer there.

He was just another old fellow living in obscurity in an unremarkable shack, until at length he became ill and was taken to the hospital. Because he seemingly had no resources he was put down as a county welfare patient, a pauper. This worried him and before he died he confided to hospital attendants that he had plenty of money with which to pay his last expenses. He revealed that he had inherited $10,000 and not trusting banks had put it into a tin coffee can and hidden it inside a stump about fifty feet from his cabin.

Immediately upon the man's death a party went out from the hospital and searched the area around the cabin but found nothing. A sheriff's posse is also said to have searched, which is logical since the county would have a lien on any money found for the amount of the man's medical and burial expenses. But the sheriff's men had no luck either. Since then hunting for the Coxcomb Hill Cache has been a popular Sunday pastime for Astorians, and hundreds of people have joined the chase at one time or another.

Such a definite location given as inside a stump fifty feet

from a known cabin should have been easy to find. Every stump on the hillside might have been chopped away until the cache was located. If the old man actually gave such clear directions, perhaps some individual from the hospital rushed out ahead of the group and picked up the cache. So if the tin can was really hidden in a nearby stump it is probably long since gone.

But there is an equal possibility that the old man's instructions were not nearly so definite as the legend indicates. If he were so ill his mind was wandering he may have given wrong, vague or misleading clues.

Hampering the search today is the fact that the hillside is greatly changed. The ravine that once plunged downward to the right of the road became the site of Astoria's second city dump. (The first dump was located along Park Road across from the city park, also on the edge of Coxcomb Hill.) When the ravine became filled from dumping, a dirt topping was put on and the site abandoned for a new dump. Some people say the fill dirt was taken from around the spot where the cabin used to sit, and the can, with cache intact, might have been thrown on top of the old dump along with the dirt.

The old cabin has crumbled to nothing and the hillside grown up to underbrush. The chances of finding the cache are practically nil, but should it have survived the early searches, and should it have survived the dirt-moving days, that coffee can with its $10,000 may still be there, and not until the lid is off will the hoary old mystery of the Coxcomb Hill Cache be solved.

Northwest Oregon

FLAVEL HOUSE CACHE

The Columbia River bar, now somewhat tamed, was in early days an extremely capricious barrier—shifting and difficult, and often completely unnavigable in times of storm. So the role of the bar pilot in bringing vessels safely into the river channel was a crucial one and many of the earliest bar pilots grew rich.

Captain George Flavel first visited Astoria in 1849 as master of the brig *John Petty*. Returning as mate and pilot of the steamship *Goldhunter* in 1850 he was issued the first river pilot license in Oregon.

The captain possessed an astute business sense as well as a sharp eye to the main chance. In San Francisco he purchased the pilot schooner, *California,* and also a tugboat or two, and he waged such bitter warfare against the rival pilot boat, the *Mary Taylor,* and Paul Corno's pioneer tug, the *Rabboni,* that both lost money and retired from service. This left Captain Flavel the undisputed entrepreneur of pilotage and towage.

For a few years he continued to command his own vessels in the coasting trade, but his pilot boat garnered such big profits that in 1858 he left the sea to manage his local investments. With the tug *Astoria,* which he built, he earned a bonus of $30,000 offered by the state to anyone who would maintain a steam tug at the Columbia River mouth for five years—and for the next two decades Captain Flavel was able to keep complete control over pilotage.

A public controversy raged as to whether Captain Flavel's monopoly because of the excellent service he maintained was for the good, or because of the high rates he charged for the bad. But one thing brooked no argument: His tightly-held control amassed him a fortune. Reputedly he is "the only man to make a million in Astoria on his own responsibility."

Early in his career he bought land across Young's Bay at Tansy Point. This was later laid out as a town planned to rival Astoria. Though the captain never lived there it was named Flavel in his honor. A fine hotel, a long wharf and other improvements were built, but the project failed and the hotel and dock fell into ruin. Though the name Flavel still appears on some maps the old townsite, now vacated, has become a part of Warrenton.

About 1857 Captain Flavel married Christina Boelling, the daughter of a white pioneer of 1847. In the 1880s he built in Astoria the many-roomed, turreted Victorian mansion which became known as Flavel House.

The captain lived in Flavel House for about ten years prior to his death in 1893 and his family continued to occupy it for forty years more. And sometime during the decade of the captain's occupancy he is said to have hidden away a cache of money —perhaps in one of the many fireplaces, perhaps in the walls, possibly even outside somewhere on the grounds. And some people say that his ghost goes stalking through the upstairs rooms on nights of wind and rain, though whether in concern over his cached gold or in the unease of a restless, troubled spirit none can tell.

The story of Captain Flavel reveals surprising points of similarity to familiar tales told of Captain James Johnson and Captain James Scarborough. The latter were both early bar pilots, settled near the river mouth (though on the north bank), made money, built houses—and left caches of money behind them when they died. Yet there are important points of difference, too. Both Captain Johnson and Captain Scarborough married Indian women, which made them not quite respectable, and both died in the 1850s. (See Hult, *Lost Mines,* "The Captains' Caches," pages 133-146, and the follow-up story in this book entitled "A Disappearing Diary.")

Captain Flavel not only married well but for forty years remained a widely known figure in maritime circles at the mouth of the Columbia, and in the city of Astoria he was a pillar of the community hardly less commanding than the Astoria Column built after his death.

The captain's daughter Nellie died in 1933. Her grandniece then deeded the old mansion to the county in the hope that its spacious Victorian beauty might be preserved. Happily this has been done, with the Flavel House now serving as headquarters and museum for the Clatsop County Historical Society.

In the interests of truth it must be admitted that Burnby M. Bell, late historian at Fort Clatsop National Memorial and corresponding secretary of the Clatsop County Historical Society, has denied there could be any substance to the story of a cache at Flavel House. His grandfather, Samuel T. McKeen, was associated in business with Captain Flavel for thirty years, his mother was a close friend to members of the Flavel family; he himself knew Nellie Flavel well. His knowledge of the Flavel family and of Flavel House had extended back over a period of forty-five years, and yet neither he nor his mother had ever heard a whisper of such a story.

He aso pointed out that the captain was a canny businessman and a millionaire who would have had no reason to hide away such a deposit of money, as he was expecting no rainy days.

According to one story afloat, a large contingent of Astoria school youths combed the big house a few years ago, after it had become a museum, but did not find any clues to the old mystery. Bell has denied that such a treasure hunt ever took place. None of the officers of the Historical Society has heard of such an episode and no such search would have been permitted.

Still, even though Captain Flavel was rich he might have had some purely personal, experimental, whimsical or eccentric reason for secreting a sum of money. In such a case he may not have wished to confide in friends or family. And certainly it seems that not all the nooks and crannies in the many rooms have been searched.

The origins of the Flavel House tale are unknown and the story itself is probably fanciful and apocryphal. But it does add a further romantic dimension to that already romantic old mansion, making it a repository not only for historic artifacts but for a lively legend of gold and ghost.

Northwest Oregon

FISHHAWK FALLS AND OTHER LOST MINES OF CLATSOP COUNTY

If you take Highway No. 202 southeast from Astoria you will find it running through rich bottoms grown up to thick stands of red alder, beautifully gray-barked and dappled like birch. Leaving the little town of Olney the road follows the Klaskanine River up the Glaser Grade and passes over Tidewater Summit. As it descends into the Nehalem River Valley it parallels the North Fork of Fishhawk Creek, where some six miles before you reach Jewell you will see to the right, not more than fifteen feet from the road, a picturesque little waterfall, Fishhawk Falls.

This country is a part of the Oregon coast range, rugged and thick grown to timber, difficult to penetrate, and very dangerous because it is so easy to become thoroughly lost. So, whether you are hunter, camper or treasure seeker, be sure to bring a compass and—be careful.

And the treasure hunter should know that according to geologists and mineralogists this is *not* gold-bearing terrain. But since no one had given that disheartening word to the pioneers of Clatsop County they managed to have a few gold strikes just like other areas, creating a few lost mine stories in the process.

One of these involved Fishhawk Falls in the days when it was still in wild country where no road ran. An elderly farmer, new to the Nehalem country, was walking through the woods when he picked up several stones which later turned out to be gold nuggets. This so excited his curiosity he returned to the falls and explored the surrounding area. He found one old digging and believed it to be the hole from which the gold had come. He planned to go back and do some mining but was always too busy to make the break—and what the hole might have yielded became a tantalizing source of speculation in the Nehalem Valley.

The mystery still lingered years later when the Murrin family of Astoria used to make excursions out there. Both Estella Murrin, who became Mrs. Eastland, and her brother William Murrin remembered hearing talk about it in their youth. Later when William visited close friends at Jewell he became acquainted with the farmer who made the discovery and heard the story directly from him.

By then the Astoria-Jewell road had been built through the mountain area at great expense, so travel was much easier. William Murrin, impressed by the farmer's story, made several trips out to Fishhawk Falls, prospecting the area. Back in the woods about half a mile he found an outcropping of rock, beneath which remained evidences of an ancient digging, now collapsed and filled with debris. He thought this could be the old shaft discovered by the farmer and he planned to try it excavate it, but always there were other things more pressing and he never did any actual mining in the Fishhawk Falls area.

This lore was passed down to Estella's son, Charles Eastland, who lives in Astoria, motivating him to make several trips to Fishhawk Falls to investigate the legend. A little upstream he found an old logging road and a decaying bridge crossing the creek, and on the other side of the highway an abandoned rock quarry, but these seemed to be of later origin. He was unable to locate the collapsed prospect hole seen by his uncle. Probably all signs of it are obliterated by now.

Some seven miles south and a little west of Fishhawk Falls, over Red Bluff Ridge and down in another part of the Nehalem drainage lies Cruiser's Gulch, where in 1901, following a discovery by Sebastian Glaser, a number of lode claims were staked, the quartz supposedly bearing gold, silver and other minerals. This location is about two miles west of present day Elsie, on the line between Sections 1 and 2, T4N, R8W; a stream named Gold Creek commemorates the site.

And in the extreme southeast corner of Clatsop County, an earlier and more prolonged gold rush took place along Rock Creek. The first claim, the Gertrude, was filed May 25, 1885 by J. M. Weed, and by July 1889 the strike had grown to such proportions that the Rock Creek Mining District was formed, with William H. Branden as president and secretary. The district comprised not only the part of Rock Creek which flows through Clatsop County but extended to its mouth on the Nehalem in Columbia County near Vernonia. Such claims as the Protector, Defender, Last Chance, Elkhorn, Mountain, Rolling and Bonanza were filed, until the whole thing petered out in 1894.

But one student of lost mine lore believes the Rock Creek region could be the true location of the fabled Lost Tillamook Mine.

Moving west to the Oregon coast, there is a lost mine in the area northeast of Seaside, though in this case the story is admittedly vague. In the 1890s an old fellow had a cabin out on the edge of Clatsop Plains. He was something of a hermit, always secretive and suspicious, and though he came into Seaside now and again for supplies he had little to do with anyone. No one in town knew or cared where he lived or what he did.

He arrived in Seaside one last time, very ill. He was taken to an infirmary where, before he died, he babbled about his mine, and in his pockets were found samples of gold dust and nuggets. The search began for the old man's cabin and the Clatsop Plains Mine, as it came to be called.

The general location is somewhere along the Seaside end of the Lewis and Clark Road, the old road between Seaside and Astoria.

And coming round robin back to the Columbia River we find that there have long been hopes of mineral wealth in this area too—mainly of magnetite, an iron oxide, though there has also been talk of ilmenite and zirconium. The black sands along the river do carry a considerable amount of iron, and in 1944 the U. S. Bureau of Mines mapped two magnetite deposits, one off Hammond on the Oregon shore, the other off McGowan on the Washington shore.

An article in the *Daily Astorian,* August 15, 1961, points out an interesting peculiarity in connection with these deposits:

"The magnetic declination of the compass in this area is about 22 degrees 30 minutes to the east of true north. A line drawn from Hammond to McGowan across the Columbia is almost 22 degrees 30 minutes east of true north: The deposits are lined up along the magnetic north-south line."

The surveys sounded so promising that the Bunker Hill Company of Kellogg, Idaho, leased the lands from the state and federal governments, but when investigation proved the sands not rich enough in magnetite and other minerals for economic production the company dropped its options.

So the dream died but tales of old lost mines and hopes of new discoveries continue to haunt the rough but beautiful Oregon corner which is Clatsop County.

LOST GOLD STRIKES OF THE OLYMPIC MOUNTAINS

In the far northwest corner of Washington State the Olympic Peninsula lifts from a rugged, sea-girt shore through magnificent spruce forests and through rarer rainforests to an interior plateau jumbled with fifty snow peaks and climaxed by 8,150-foot Mount Olympus with its three summits.

The unique features of the Peninsula, which is "one of the most primitive and untouched areas of beautiful mountain scenery in the United States," have been preserved in the 890,000-acre Olympic National Park, home of the Olympic elk.

In the beginning there were great hopes that the area would yield mineral wealth as fabulous as its beauty. It was claimed that streams had shown colors and that "large, well-washed nuggets" had come out of the mountains on the Gray's Harbor (southwest) side. In 1877, before any important mineral discoveries had been made on the Peninsula, one newspaper attributed "valuable deposits of gold, silver, cinnabar, copper, lead, iron" to the Olympic Mountains.

A miner just returned from the gold fields of New Zealand said, as with prophetic wisdom: "No one can assert that it (the Peninsula) is not as rich as the richest gold fields in Australia or Cassiar." And all awaited the strike that would prove the point.

Hope could continue as long as the mountain area remained shadowy and unknown. And though white settlement was made around the perimeter of the Peninsula in the early 1850s, the inexplicable fact is that the interior lay unexplored for nearly forty years. A mysterious aura surrounded the Olympic Mountains and speculation arose as to what would be found within their fastnesses. The talk was of a great central valley, wide-reaching prairies, lost Indian tribes, a lake with a subterranean outlet—and fields of gold.

In the winter of 1889-1890 the mountains were at last penetrated by an expedition financed by the Seattle *Press* and led by James H. Christie; in the summer of 1890 the whole interior area was thoroughly examined by different parties sent out by an Army expedition in charge of Lieutenant Joseph P. O'Neil. And what was discovered was simply mountains and more mountains, "with views unequaled in the world." The exploring parties brought out no encouraging evidence of a mineral nature.

But miners remained undaunted, and in the late 1890s the more accessible eastern slopes of the mountains back from Hood Canal became the scene of intensive prospecting and digging as men searched for gold, silver, copper, lead, iron, manganese. On Iron Mountain in the upper Quilcene area some promising strikes were made in copper, and on the copper claims named Tull City, Tubal Cain and Copper City shafts were sunk. But in the end the ore proved not rich enough to pay its way out of the mountains.

Yet the belief persisted that rich mining strikes lay just "over the hill" somewhere on the Peninsula, and in 1909 when President Theodore Roosevelt set aside the central mountain area as the Mount Olympus National Monument to save the Olympic elk from extinction, a great howl went up from mining men who swore the area included great potential mineral wealth. Their lobbying was so loud and strident that in 1915 President Woodrow Wilson cut the Monument in half—eliminating, according to the U. S. Forest Service, the part which gave promise of mineral discoveries, especially manganese needed in World War I; but the only thing "mined" was the timber reserves released to hungry lumbermen.

There was so little in the way of mining activity in the next twenty years that in the big fight during the 1930s over the question of a national park, the main opponents to a large park were the timber interests, with almost nothing heard from mining men.

And so though the Olympic Peninsula has many claims to fame, among them is no claim to mineral wealth. And yet the old dream still persists and the tales told and retold make one pause to wonder: Could the secret of a great strike be guarded by the mountain god of the Indians, the Thunderbird, who dwells on top of the highest peak, making lightning by a flash of his eyes, making thunder by a flapping of his wings?

It is perhaps for the best that no gold fields were discovered, because evidence of mineral deposits would have hindered the establishment of the Olympic National Park. The park itself represents great riches of another kind: sweet, resinous forests, gem-like lakes, high, wild mountain basins, flowered alpine meadows, ranges of snow peaks, unique rainforests, herds of Olympic elk, and vistas of almost heartbreaking beauty—not "locked up" as its enemies are wont to say but preserved for all the citizens of our country. Everyone should visit it, walk its superb trails while renewing himself in the vastness of the unspoiled landscape and breathing the mountain-spiced air.

The tales which follow add another, perhaps all-too-human dimension to the scene. And while these lost prospects may be sought simply out of a desire for romance and adventure, it must be remembered that should they be found within the boundaries of the Olympic National Park no mining will be permitted. Surrounding the park, however, lies the Olympic National Forest where the land is open to mining claims.

Lure of the Lost Placer Deposit

This story dates from the time when the copper strikes in the upper Quilcene Valley were making a stir in mining circles, in the decade between 1895 and 1905. A party sent out by "Senator Clark" was in the field—which must mean Senator William Clark of Montana, one of the copper kings of Butte. Although he had become a millionaire through his exploitation of "the richest hill on earth," he was still not averse to finding new copper mines and making new millions. But his Olympic Mountain prospecting venture seems to have gained him nothing of value in the way of copper leads. It is remembered for a different matter altogether.

A member of the party leaned over the stream to drink one day and discovered gold in the gravel. Either the man quit the Clark party or else waited until the job was finished. He then returned to that high Olympics stream to work his prospect. After making a rough sluice box from cedar shakes cut from a dead tree just below timberline, the man bent his efforts to mining and took out gold which he later sold in a Seattle assay office for $600.

He worked his digging for several summers. Before he died around 1910 or 1912 he passed his secret on to a man named Noe in Montesano, Washington, who in turn passed it on to a prospector, C. R. Terrell.

Terrell, becoming convinced he could find the placer location, invaded the Olympic Mountains on the lost mine search in 1914. Not having the quick luck he had envisioned he spent four consecutive summer seasons in the mountains. Sometimes he took with him as a partner Bill Burrows of Montesano. Once Terrell wandered around lost for four days at the head of the Skokomish River and then ran into a fire crew which pressed him into service as a fire fighter. In camp Burrows nearly went wild with worry.

In the fall of 1917, after the United States had entered World War I, Terrell was forced to give up the hunt.

Many years later, after Terrell had become a desert rat down around Tonopah, Nevada, he told the story of the lost placer deposit to a newspaperman, revealing that he still believed that with a little more time he might have had luck in locating the old cedar-slab sluice box. Even then he dreamed of returning to the Peninsula for one more try.

"On my last trip to that land of no end of canyons," he said, " . . . . I got pannings of coarse gold high up in the Olympics but I am not saying which river it was . . . . I'm going back there to take up the trail of the placer mine where I left off."

The Mountain Peak of Gold

Starting out from Brinnon, Washington, on Hood Canal, a hunter with a horse and pack outfit followed an old government trail into the Olympic Mountains. This must have been the trail blazed by Lieutenant Joseph P. O'Neil's party in July 1890 from Lake Cushman up the Skokomish river valley as far as the headwaters of the North Fork.

After the hunter had traveled some distance into the mountains he shot a deer and wounded it. Tying his horse to a tree he started out on foot, followed the trail of blood, and was led on a long chase. He did not notice that the afternoon light was failing, until fog came suddenly down, enveloping all landmarks. He was lost. Only then did he realize how foolish he had been, for he had left food, blankets, compass and everything else he needed strapped to the horse. He was deathly afraid of mountain lions, and in a panic he went stumbling uphill and down all night. He never did find his horse, and as the days went by he would take shelter where he could, and having his rifle he was able to subsist on game.

Somewhere in his wanderings – and he did not know where, because the whole experience was later one ghastly blur—he came upon "several human skeletons, and on top of the mountain he found the ground covered with gold and tossed some of it down the hillside to see it roll."

Stuffing his pockets with what he could carry he pushed on, and after terrible hardships and exposure, came out on the ocean side of the Peninsula to a farmhouse. His hair had turned white by the fright of his experience.

The specimens he brought out in his pockets turned out to be $800 worth of almost pure gold. He did not care. He wouldn't go back to those deadly mountains for all the gold in the world.

But the story passed on to an old Arizona prospector, Richard Stevens. The wilds held no fear for him and he decided to try to locate the peak from which the gold had come. A friend of Stevens who had been a trapper and prospector in the Olympic Range had come to the conclusion that most of the gold lay on the north slopes of the mountains. The trapper had talked with an Indian who had a supply of gold and said he had found it in a small creek on the north side.

From this it is believed that Stevens, who spent a number of summers in the Olympics, conducted his search on the north slopes. Years later, after he had become an old man living in retirement in East Stanwood, Washington, he reminisced about his trips to the Olympics but admitted that he did not find the peak of gold.

In assessing this story it must be admitted that a hunter could get hopelessly lost in the Olympics under the circumstances cited. His panicky blunderings about sound very real in the telling. But what of the skeletons he claims to have found on the mountain top? Could that be true?

Yes, if in his state of anxiety one skeleton was magnified to look like several. For a human skeleton certainly must lie somewhere in the northeastern part of the Olympic Mountains.

Lieutenant O'Neil, leader of the successful Army expedition of 1890, made an earlier, abortive attempt to penetrate the mountains. In 1885 he started from Port Angeles with a party of men. Crossing the upper valley of the Dungeness River, he set up base camp in rough country a little to the southeast. Taking two men he then explored a ridge which ran southwest above the East Fork of the Elwha River, and continued on toward Mount Constance.

During the return to base camp, Private Johnson, one of O'Neil's men, lost his way somewhere along the trail and was never heard from again. Quite possibly the young private, on finding himself wandering in circles, scrambled exhausted to the top of a mountain to try to get his bearings. He is presumed to have died of exposure in the wilds, and his skeleton may be the one which guards the hunter's peak of gold.

The Cave of Gold

Elwha Sam was an old time hunter of the Clallam tribe which ranged from the north slopes of the Olympic Mountains to the shores of Juan de Fuca Strait. His home was on the banks of the Elwha River, a large stream draining the mountain area and emptying into the Strait just west of Port Angeles. In those days Port Angeles was a mere village.

Sam started out one October day to secure meat for curing against the winter weather which was fast approaching. Having hunted elk and deer in this part of the country all his life, he paid little attention to landmarks as he followed the Elwha upstream and then struck out into the trackless mountains. But signs of game proved sparse and he ranged farther than was his usual custom. Twilight caught him far from the shelter where he had planned to spend the night. He was tired, hungry and cold.

The darkness of a storm closed in suddenly and rain began to pour. From around him in the eerie woods came "the hoot of owls like evil spirits."

All the superstitions of his Indian life rushed up to frighten

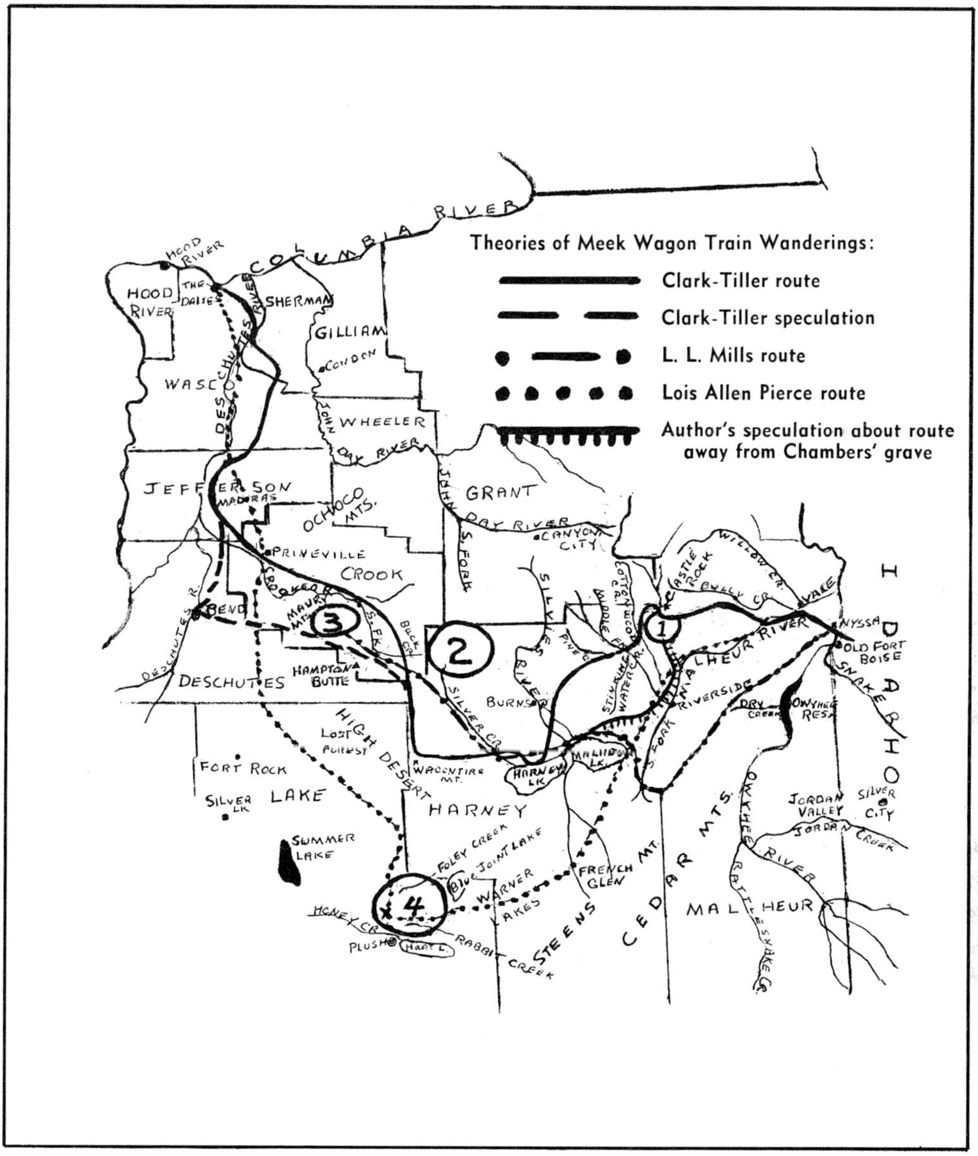

EASTERN OREGON

1—Site of Mrs. Chambers' grave

2—Stone huts, upper Silver Creek

3—Site of Mills' "discovery" of Blue Bucket Mine

4—Site of mine, according to Mrs. Pierce's theory

Lowell Tiller and Keith Clark, authors of *Terrible Trail: the Meek Cutoff, 1845*, are shown examining a mule shoe found along a well-worn trail near Stinking Water Mountain in the Malheur country.

Left: "Uncle Billy" Bartlett, well-known character of the Orofino area, who sought the Lost Indian Post Office Mine. (Blanche Casteel) *Right:* Don C. Tracy of Lakeview, Oregon, who confounded the experts when he discovered Oregon's first (and still richest) uranium mine, in Lake County.

him—and he panicked. Below him he heard a murmur of water, and believing the sound to be that of the river whose banks he could follow away from this haunted place he started blindly downward, slipping and sliding as he went. He had fallen and was crawling on his hands and knees when he felt a dry spot beneath him. He realized he was under a ledge, which gave him a sense of security, and he edged in closer to a rock wall he could feel before him, trying to get in out of the rain. Suddenly he pitched forward.

When he awoke he was lying in a cave, on hard rocks, his whole body stiff and sore. As he opened his eyes he was aware of a glittering in the dark. He feared some kind of spirit magic and hurried to strike some matches he carried in his pocket.

"Listen. It was not magic, but gold, yes hyu chicamin; gold. What I tell is true but no white man can find the cave. It is hidden by brush and weeds and where it is I shall never tell. Inside there is a steep bank, and that is where I fell from . . . .

"Many years have passed since I found the cave of yellow chicamin. The Great Spirit of the caves and mountains and the rushing Elwha led me to it but he keeps it hidden from others."

Such are the words of Elwha Sam as set down by that talented daughter of a Port Angeles pioneer family, Mary Gay Morse, in her charming, slender book, *Lore of Olympic-Land*.

Mary Gay's was an imaginative and sympathetic nature. Always she was fascinated by the stories and legends indigenous to that mountain land she loved so well. To her the greatest riches in tradition and lore belonged to the Indians of the Peninsula. She felt that because the native people were so gifted with sharp discernment and primitive insight nature might reveal to them secrets carefully hidden from crass white men. For this reason the story of the Cave of Gold might well be true. Even the white pioneers of Port Angeles who heard Elwha Sam tell his story accepted it, she said, "not without credence."

She concluded: "The silvery waters of the Elwha flow placidly on, and rugged and silent the mountain peaks stand . . . . but the forest . . . . has wrapped its mysterious arms around the unknown spot, and only time will tell whether the tale of the Cave of Gold was a fact or mere tradition."

Cascade Mountains, Washington

LOST MINES OF THE MOUNT ADAMS AREA

Of the many beautiful features natural to the Pacific Northwest perhaps the most awesome are the five volcanic cones which lift their ethereal shapes along the Cascade Mountain range in Oregon and Washington. Four of these, Mount Baker, Mount Rainier, Mount St. Helens and Mount Hood, are located just west of the main range and are visible for long distances on the mild, rainy side of the mountains. Lesser known is Mount Adams because it strides the main range where it trends east almost to the Yakima and Klickitat valleys and so lies in isolation away from population centers and main traveled roads.

Among the five incomparable peaks Mount Adams is second in mass and height only to Mount Rainier. It is indeed a mighty mountain with four distinct summits in the clutch of nine named glaciers. Exposed to the long sweep of the western wind it towers in lonely majesty above the low ranges which surround it. Secondary volcanic cones are known, and picturesque crags of yellow, coral, violet and vermillion protrude from the slopes at snow line.

Oddly enough, Mount Adams' very summit was the scene of mining activity in the 1930s after it was discovered that a sulphur deposit of hundreds of tons lay under the icecap. Interested men formed a company and filed a claim. They brought a model-T engine up the mountain to run a drill and then used pack horses to carry bulging loads of sulphur down the easy south slope.

This operation came to an end in 1937. To protect the great mountain from such despoliation, as well as to preserve adjacent forests, meadows, lakes and streams, the U. S. Forest Service in 1942 set aside the 42,411-acre Mount Adams Wilderness, now a part of the National Wilderness Preservation System.

In the Wilderness the only mining which can be done is by hand, with ore packed out by back or by mule, no mechanized equipment or vehicles being allowed. So if either of the two lost mine stories which follow leads to locations within the

Mount Adams Wilderness, searchers will be restricted to the primitive methods of locomotion and mining used by the original discoverers—which should prove a challenge.

But in both cases it is more likely that the old mines lie in the surrounding Gifford Pinchot National Forest, where mining claims are permitted.

Rabado's Lost Gold Mine

In the late 1880s Pierre Rabado, living near Bickleton in Klickitat County, Washington, used to disappear into the Mount Adams area and return a few weeks later weighted down with as much as $500 worth of gold dust. He was close-mouthed, managed to evade the curious and so kept his secret well. No clues pinpointed the location but it was generally believed he procured the gold somewhere along the western base of Mount Adams.

In the spring of 1891 Rabado left for his usual mining trip but this time failed to return to Bickleton. Whether he died of natural causes in the wilds, was killed in an accident, was murdered by the Indians, or went out of the mountains by some other route was never determined.

After the Frenchman's disappearance, Amos White, a miner and rancher in the White Salmon River valley, which lies south of Mount Adams, pursued information about the lost mine. He once showed pieces of gold quartz he claimed came from the mine and assayed $3000 to the ton. How he came by such rich specimens when he did not know the location of the mine is not known. But with the quartz in hand he was wild to ferret out Rabado's secret. He gained the confidence of a Yakima Indian who reputedly knew the mine location and offered the Indian $500 as a reward plus $1000 a year income for life if only the Indian would lead him there. Or such is the story told; if there is any truth in it Amos White must have believed the mine fabulously rich.

The Indian accepted the offer and the two left for the mountains. They had been absent about three weeks when they met a Yakima medicine man who threatened White's guide with such dire punishment if he should reveal the mine location that he became afraid and refused to cooperate further. But White had picked up enough clues from the Indian that he later placed the mine twenty miles northwest of Trout Lake, which is a little settlement southwest of Mount Adams.

White insisted he could lead a search party to Rabado's Lost Mine but the country where it lay being so rough and uncharted he would need both money and companions. Early in 1909 he formed the Mount Adams Prospecting Association, a stock company with members drawn from the farmers of the White Salmon Valley.

In July of that same year White and four association members left Husum with a pack train carrying provisions for three weeks and headed up the White Salmon Valley and over the divide into the Lewis River drainage. After an exhausting trip of several weeks, the men returned unshaven, unwashed, and outspoken in their disgust—and with White bringing up the rear. As soon as they had entered the wilds, the men said, White's certainty had fallen away and he had blundered about in erratic fashion. Finally they had forced him to admit he did not know the way to the mine.

White's spectacular failure lost him the confidence of his backers and the prospecting association was disbanded.

The amazing thing about this story is how many of the details dovetail with those of the Lost Spaniard Mine (Hult, *Lost Mines,* pages 149-152). For the Lost Spaniard Mine, too, was thought to lie "somewhere at the head of the Lewis River, in the wild, almost inaccessible territory between St. Helens and Mount Adams," and according to legend it too yielded rich gold in the 1880s, which the Spaniard brought out by pack train, passing through the Klickitat Valley to take his money to a bank at The Dalles.

Best evidence indicates that the Spaniard, too, lived in the Klickitat Valley, where the two foreigners, being strangers in a strange land, might naturally have become friends. In this case the secret of the mine passed from the Spaniard, its discoverer, to the Frenchman, who would seem to have begun his pilgrimages only after the Spaniard's death.

In both lost mine stories Yakima Indians reputedly knew the secret of the mine location; the secret of the Lost Spaniard Mine was offered for sale by a Yakima tribesman as late as 1922.

Lost Mines of the Mount Adams Area

In 1890 the mountaineer, C. E. Rusk, for whom Rusk Glacier on Mount Adams is named, circled the great mountain on horseback, traveling most of the way on well-defined Indian trails which laced the area. He told of large contingents of Indians invading the area northwest of Mount Adams to pick the huckleberries which grew there in extensive patches.

To the Indians this was their hunting grounds, their berry fields, their land to roam. And if certain Yakima Indians had discovered a source of gold somewhere beyond Mount Adams, this too was theirs—and their resentment over white men's invasion may have resulted in the death of the Spaniard, and after him, of Rabado, the Frenchman.

If all this proves, as it tends to, that the Lost Spaniard Mine and Rabado's Lost Mine are one and the same, then we must combine the clues from each of the two stories. We might then look twenty miles northwest of Trout Lake (as Amos White believed) for a certain waterfall (a legend in connection with the Spaniard's mine), behind which should lie a cavern filled with gold.

The Cave Of The Hidden Mine

In 1937 Frank Smits of Olympia, Washington, was developing a quicksilver mine in the Mount Adams area. Needing a shorter route for hauling in his supplies he went exploring along terrain he had never investigated before and at the base of a cliff noticed a hole leading underground. Deciding to investigate he lit a torch and crawled in—to discover a vast subterranean vault, which he followed for a long way uphill, until he came to a drop-off, with an old wooden ladder leading to lower depths. Descending he found himself in a smaller vault, through which ran an underground stream, besides which sat an ancient sluice box.

Shining his light in different directions he saw a corner of the cavern had been used as living quarters by the miner. An old stove stood mutely, and other such items lay around as pots, pans, a shovel, an axe, a muzzle loading gun, and a pistol, all covered with rusts of time. And on a crude bed near by was stretched the skeleton of what looked to have been a large man.

An old trunk leaned against the wall of the vault, and in an effort to solve the mystery he had stumbled upon, Smits opened it and rummaged through its contents. He found a Bible and other musty books, a tintype picture of a woman and children, and some papers from the State of Virginia, dated 1878. There was also a legal paper regarding a lot transfer in Tekoa, Washington Territory, February 10, 1881, bearing the name of Isaac N. Huffman.

Was this the name of the man whose skeleton had lain here undiscovered beside his mine for half a century? And what anguished hours his wife and children must have spent wondering what had become of him!

Such was the strange story that Smits told a reporter when he passed through Portland, Oregon, on his way home to Olympia.

And the more one considers it the more incredible it becomes. What circumstances could have led a man to leave sky and sunshine behind and set up his camp in a cavern two levels down? Was he an insecure type who took comfort in retreating to a caveman existence? Or was it mere expediency? Had he happened upon a rich winnowing of gold along the underground stream — wealth he hoped would speed his return to home and family? Was he ill when he made his camp down inside the dark earth — or what happened?

Most of all we would like to know if Smits found evidence of gold in the decaying sluice box, or found gold anywhere in the effects of the long-dead miner. The omission of this information seems curious, since surely a reporter would have asked Smits about it. Did Smits evade the question — or even ask the newspaperman to remain silent on the subject because he hoped to return to mine the subterranean stream?

We do not know.

Mount Adams does have its beautiful Ice Cave at 3000 feet, "a bubble in the plastic rock," as the *WPA Guide to the Evergreen State* describes it. "The entire region south of Mount Adams is full of these caves," says the *Guide,* "most of them without openings but known to exist by the hollow sound made when horses clatter over them."

And somewhere among these caves lies one reached through a hole at the base of a cliff which shelters the long-hidden mine of the mysterious man who left his skeleton to moulder in the eternal silence of the underground.

Cascade Mountains, Washington

CHIEF KITSAP'S GOLD MINE

During the time that Indians and whites were still living in partial tranquility in what is now western Washington, a chief named Kitsap mined gold in the back reaches of the Cascade Mountains. According to legend his mine lay somewhere at the headwaters of the Greenwater River, east of Mount Rainier.

But there was more than one Chief Kitsap and they have become confused in white men's histories. Elwood Evans mentions a war chief and medicine man of that name "of the band whose haunts were upon the Peninsula adjacent to Port Madison." Theodore O. William wrote of the same or another Kitsap, "He was, from 1770 to 1845 the most powerful chief that ever the Indians of Puget Sound saw" — this because about 1825 he banded all the Sound Indians into a large confederation, the better to raid Vancouver Island. William calls him " the greatest warrior of the last century and the greatest clam eater of them all."

When the Indian War of 1855-56 broke out in western Washington a Chief Kitsap joined the hostiles and became a leader. Bancroft calls him "the younger," and it is natural to assume he was the son of the great Puget Sound warrior. Instead his origins were on the east side of the mountains rather than on Puget Sound. He was a Klickitat Indian, said to have been a brother of Leschi and Quiemuth, who became chiefs of the Nisqually tribe; he himself became a chief among the Green and Muckleshoot tribes. Possibly this Klickitat adventurer adopted the name Kitsap simply because it was an illustrious one.

When the war went against the Indians Quiemuth was stabbed by a white assassin, Leschi hanged, and Kitsap arrested but later released. In 1858 he was reported making trouble on the Green River "with a small body guard, tolerably well armed." In jail Kitsap had been introduced to white man's red tonic, and being a medicine man he later administered some red paint believing it to be the same material. He lost some patients and in 1860 was killed by one of his own people in retaliation.

Kitsap County, of which Bremerton is the largest city, is often said to have been named for this hostile chief but it would be much more logical if the name referred back to the great clam eater who lived and ruled in the area.

As for the lost mine story, it was undoubtedly the Chief Kitsap who was connected with the Green River and Muckleshoot Indians who brought the gold out of the Cascades. For only a few miles separates the tributaries of the Green from the Greenwater, and these mountains were the natural range of the Indian bands which Kitsap ruled.

As the tale is told, the chief discovered a wide vein of gold along a steep bluff on the side of a canyon just below snow line. All he had to do was knock off and "gather nuggets the size of robins' eggs" and on down to the size of navy beans. He could also pick gold nuggets out of the creek bed below the bluff.

Once he took his wife to the mine and with her an Indian slave woman. The adventure so impressed the Indian slave woman that she retained in her memory many incidents of the trip and the main landmarks along the way. Years later, after Kitsap was dead and the mine all but forgotten, she imparted the information to Anthony P. Carr of Tacoma. He was a kindly man who had perhaps befriended her in her old age.

As a pioneer who was one of the founders of Tacoma, Carr is well remembered on Commencement Bay. Born in Indiana in 1841 he distinguished himself during the Civil War by carrying an important dispatch from the 19th Indiana Volunteers through the forests of Hall's Hill to Washington, D.C., where he handed it to the great Abe Lincoln himself and was given new clothes and a bath at the direction of the President. A year later when Lincoln was reviewing the regiment he called Carr from the ranks and congratulated him on his trip from Hall's Hill.

On being released from the Army Carr traveled west by ox team, arriving on Puget Sound in November 1865. He was so taken with the country he immediately started back to Indiana to persuade is father, Job Carr, to emigrate. Unknowingly he passed his family en route, they already having decided to follow him west. Learning this he hurried back to Puget Sound, where he helped his father build the first log cabin on Commencement Bay. The cabin is preserved as an historic landmark in Tacoma's Point Defiance Park, where it can still be seen.

Anthony Carr bought a good deal of land within the town, and in 1869 in platting and recording the property was first to use the name Tacoma in a legal way. That same year he married Josephine Byrd and for about twenty years, until his wife's death, he was a leader in his adopted city's attempts to rival Seattle.

As a widower he retired to a farm at Spanaway, where he cultivated his acres beside the lake. Friends and neighbors described him as keen of mind and witty of speech but "of rather eccentric habits."

The charge of eccentricity no doubt resulted in part from the fact that in the 1890s he began a search for Chief Kitsap's gold mine. He followed the clues imparted to him by the Indian slave woman and after spending a number of summers in the Cascades announced that he had been successful in his quest.

"A. P. CARR OF TACOMA THINKS HE HAS FOUND THE TREASURE," read a headline in the Seattle *Post-Intelligencer*, July 28, 1900, over a special dispatch from Tacoma. Apparently his place of discovery was somewhere at the head of the Greenwater but "he will not give out any statement as to its exact location." He had also found a large vein of brown hematitic iron ore he proposed to have tested by experts.

And what happened then? Did he develop the mine? Did he visit it each summer, bringing back a season's wealth of nuggets? Or was he content once he felt he had solved the mystery? A curious blank exists on the subject.

Anthony P. Carr died in 1923 at the age of eighty-two. The Tacoma newspapers carried lengthy obituaries together with photographs showing him to have been a man with high forehead and heavy drooping mustache. He was hailed as a prominent pioneer who helped to lay the foundations of Tacoma and his life history was reviewed, all except that part when his main goal was to discover Chief Kitsap's Gold Mine.

A story related to me by a friend, R. H. McGee, is rather too vague to stand alone but may be of some interest in connection with the Kitsap story. In the 1920s he heard tell of an Indian who used to come into Mossyrock, Washington, with gold and before he died disclosed that his mine was near a place called Soldier's Rock. This was believed to lie at the headwaters of the Sispus River, a tributary to the Cowlitz, which heads in the area between Mount Adams and Mount Rainier.

Twenty years later McGee heard a similar story which placed the mine near Soldier's Peak on the Green River, or perhaps the Greenwater – the teller being not quite sure which river was meant.

So if Soldier's Rock and Soldier's Peak are one and the same place, and the Soldier's landmark is located on the Greenwater River, this tale could pertain to Chief Kitsap's Mine. Admittedly, I have found no "Soldier's" place name on the map but the landmark may be too insignificant to show—and maps are notoriously incomplete anyhow.

At any rate, a canyon supposedly yielding nuggets the size of robins' eggs on down to the size of navy beans seems well worth looking for in the shadow of that mighty white giant, Mount Rainier.

LEGENDARY LOST GOLD MINE OF THE COEUR D'ALENES

Is there a fabulous gold mine still undiscovered closer to Lake Coeur d'Alene than the gold and silver belt which has made the Coeur d'Alenes so famous as a mining district?

The story of a Lost Coeur d'Alene Gold Mine or Lost Cabin Mine was at one time a familiar topic to miners and prospectors all over the Pacific Northwest. The legend goes back to the beginnings of white man's history in the lower Coeur d'Alene Valley.

In 1846 the Jesuit missionary, Father Pierre Jean DeSmet, selected as the site of a mission to the Coeur d'Alene Indians a prominent knoll near the Coeur d'Alene River, about twenty miles east of beautiful Lake Coeur d'Alene. Completed in 1853, the Mission of the Sacred Heart served as a rendezvous for passing explorers, traders, trappers and other travelers across the old Indian trail through the Coeur d'Alene Mountains.

In 1859 Captain John Mullan of the U. S. Army was assigned the task of building a military road from Fort Benton, Montana Territory to Fort Walla Walla, Washington Territory. The Mullan Trail, as it was called, crossed into Idaho near Lookout Pass and followed the South Fork and the Coeur d'Alene River down past the Mission. From there Captain Mullan planned to put the road around the south end of Lake Coeur d'Alene but being defeated by the spring flooding of the St. Joe River he rerouted it past Wolf Lodge Bay, an east-reaching arm of the lake, and west to what is now the city of Coeur d'Alene.

During this road-building activity Captain Mullan saw the first nuggets ever found in the Coeur d'Alene Mountains. A number of years later he wrote:

". . . . I now recall quite vividly one of my herders and hunters, a man named Moise, a French-Canadian . . . . coming into camp one day with a handful of coarse gold which he said he had found on the headwaters of the north fork of the Coeur d'Alene* while out hunting for our expedition. This gold was so pure, so heavy and so free from quartz or matrix rock and weighing several ounces that it attracted not only a great deal of attention but some degree of surprise at the time. And when he said he had found it his statement was scarcely credited by anyone in our camp, because it was believed he had traded for his gold with some party passing up and down from the Fraser River mines."**

This skepticism about Moise's handful of gold seems odd since among Mullan's men were many veteran prospectors who recognized the terrain as being capable of such a yield. Captain Mullan writes: "The members of my expedition were composed very largely of old miners from California, and having had more or less experience in noticing the identations of mineral deposits, their universal verdict was that the entire country from Coeur d'Alene Lake on toward and including the east slopes of the Rocky Mountains (meaning what we call the Bitterroot Mountains) was one vast gold bearing country."**

Captain Mullan also learned that Wolf Lodge Prairie contained traces of gold, and he found a camp of prospectors in the area in 1862. But he kept all talk of gold quiet during the roadbuilding project lest his men desert to go prospecting. So no gold rush to the Coeur d'Alenes took place until 1864.

In that year a man named Charles Wilson appeared in Walla Walla much excited about a gold discovery he claimed to have made in the Coeur d'Alene Mountains. Enthusiastic followers flocked to him and he led a party to the Mission of the Sacred Heart and then on into the mountains, where he became confused by the many canyons and timbered slopes all looking alike. The stampede party became annoyed at his failure to lead them to paydirt and threatened to hang him. His friends managed to smuggle him back to the Mission, where priests found two Indians to guide him back to Walla Walla.

---

* Now designated on U. S. Forest Service maps of the Coeur d'Alene National Forest as the main Coeur d'Alene River. What was formerly the Little North Fork is now labeled the North Fork. Locally, however, the old usage of names is still maintained.

** From a letter by Captain John Mullan written to the Murray *Eagle* in 1884.

Is it possible that Charles Wilson, thankful over his deliverance, told the priests all he could remember about the way back to his diggings, and that someone from the Mission, less raddled than he, had better luck in the search? A fascinating though admittedly unsubstantiated tale has long been whispered to the effect that Father Joseph Cataldo, who served at the Coeur d'Alene Mission from 1865 to 1870, knew a source of gold along the Mullan Trail. Some say this was Charles Wilson's original bonanza and the later site of the well-known Idaho Goldfields mine, located near the summit of the Fourth of July Canyon. According to this muted gossip Father Cataldo used Indian converts as miners to gather wealth to further the work of the western missions.

We do know that the Wilson story inspired a small rush to the Coeur d'Alenes in 1865 and that digging was done, according to a Portland *Oregonian* item, "at no great distance from the Old Mission." If some of these stampeders spilled over into the St. Joe Valley, as would be natural, credence would be lent to another shadowy story, that of the Lost St. Joe Gold Mine, which dates from around 1865.

It is said that two prospectors came down the St. Joe and then ascended the Coeur d'Alene River to the Mission, where they traded for supplies with gold. The temporary release from their lonely life loosened their tongues and they confided that their mine was on the fifth creek up the St. Joe River from the confluence of the St. Maries River. On their way back to their camp they were killed (drowned? attacked by Indians?), and the general location of their gold source being known Father Cataldo is believed to have sent his converts to work the St. Joe mine, five creeks above the St. Maries branch.

The St. Joe Valley came to be rumored so rich in gold that a gold rush in 1868 lured men all the way from California, one of their number going on to discover the Cedar Creek mines just over the Bitterroot Divide in Montana. (See the story, "Chinee Cache of China Gulch.")

In the early 1870s a well-equipped and serious prospecting party entered the mountains, the story later being recounted by John P. Vollmer, an important merchant and railroad man of central Idaho. In Lewiston his mining interest was first sparked in 1872 when a friend, George A. Frost, returned from a brief summer excursion through the Coeur d'Alene and St. Joe mountains with a firm conviction that the northern region was rich in mineral potential. Frost and Vollmer persuaded Frank E. Peck of Boston and three other men to join them in an extensive exploration. The party of six left Lewiston in June 1873 and after visiting Montana began what Vollmer's account describes as "a thorough investigation of the St. Joe on the north and the Clearwater on the south, including nearly all the parts of the Coeur d'Alene that had not been prospected."

Actually one summer would not have sufficed to make a thorough mineral examination of even a small part of any one of the three vast areas he mentions. To have paid even cursory visits to the St. Joe, Clearwater and Coeur d'Alene drainages in one short season they would have had to keep to main trails and allow time only for brief side prospecting trips.

They found one ledge so promising they chipped off a packsack full of rock which they lugged back to Lewiston. When assays showed a value of $1500 in gold to the ton they were much impressed. In the spring of 1874 they even organized and named their mining district – it was called the Agawam – before setting out to return to the ledge and begin to open it up. But like Charles Wilson they found all the canyons and ridges looking alike and they were unable to locate their last year's prospect. They returned to Lewiston highly disgruntled, leaving the tale of their lost ledge to reverberate across the North Idaho Panhandle.

Vollmer unfortunately gives no clues as to the general area in which they found and lost the prospect but most surely they entered the Coeur d'Alene country by the only means of easy access, the Mullan Trail. A suggestion as to where their strike might have been made lies in a story still afloat in central Idaho called the Lost Gold Mine in Fourth of July Canyon. Grace Jordan in her book *Home Below Hell's Canyon* tells of the obsessive interest in this lost mine story of a friend and neighbor, Charles A. Russell, and the complications of his having to consider the existence of four Fourth of July canyons in Idaho.

In point of formal fact there are four Fourth of July creeks but only one Fourth of July Canyon. The four creeks all lie in the Salmon River drainage. One flows down from the Bitterroot Divide to join the main Salmon a little south of the town of North Fork; a second is a tributary to Panther Creek — these two within the Salmon National Forest. A third forms a part of the Yankee Fork drainage in the Challis National Forest. A fourth lies farther south, joining the upper Salmon near the Sawtooth Valley District Ranger Station in the Sawtooth National Forest. Though these mountain streams can be said to pass through canyons the name is attached to the creeks and not to the surrounding terrain.

In the Coeur d'Alene Mountains the opposite it true: there is no Fourth of July creek but ever since 1860 when Captain John Mullan and his men celebrated Independence Day in the long defile which leads west from the Old Mission through the mountains this section has been known prominently as the Fourth of July Canyon. The top of the pass is called the Fourth of July Summit. The trunk of a pine on which Mullan's men carved the date of their observance is still to be seen and is frequently visited.

Though Mrs. Jordan brands the story of her neighbor's interest as "one grand fairytale" she reveals that when Russell made a trip out in search of the "incredible vein" he traveled *north.* All the Fourth of July creeks lay south. Yet the version which Russell heard was not the Vollmer story but involves a man who while crouching along a stream where he had taken refuge during a forest fire spotted a deposit of rich gold quartz, but the ravages of the fire so changed the look of the country he could not orient himself when he returned.

Still, the two stories tie themselves at least loosely together.

No discoveries of importance were made in the Coeur d'Alene mountains until the 1880s. In 1883 A. J. Prichard made a spectacular gold strike on what was afterwards known as Prichard Creek, a tributary to the North Fork (now the upper Coeur d'Alene River). In 1885 Noah Kellogg while looking for his strayed jackass (or so the story runs) discovered the first of the lead-and-silver veins of the South Fork which have since yielded staggering wealth.

Yet neither of these never-to-be-forgotten strikes answered the description of the Lost Coeur d'Alene Mine or Lost Cabin Mine.

The Prichard Creek discovery was made too far from the lake (about forty air miles or sixty by way of the river), and the South Fork discovery was of glittering galena.

On Prichard Creek the hectic boom towns of Eagle City and Murray flourished after 1883 – but only for a few brief years, then rapidly waned. Gold miners left at loose ends turned their minds back to the legend – for if such a rich gold field lay in the upper Coeur d'Alene country, why shouldn't an equally productive one be found in the lower valley? Prospectors began giving closer attention to the country near Lake Coeur d'Alene.

In 1891 the first two claims were staked just south of Beauty Bay which gives Mineral Ridge its name. Many other claims were taken and much poking and prodding done as locators dreamed of making a fortune on "The Ridge."

But to some a mine was not enough: they wanted to find the Lost Cabin. Such was the object of search during the summer of 1896 of a man named Abrams and his wife. One day, in exploring a steep canyon they separated, the wife taking one side, the husband the other.

Mrs. Abrams toiled up an ascent so rugged that the hardiest prospector might have been stopped. Over her head the pine trees towered nearly 200 feet in height, obscuring everything in the gloom of their heavy growth. About two-thirds of the way up the canyon Mrs. Abrams came upon a log cabin–deserted but giving the appearance of having been lived in by white men. Not far from its door the dark mouth of a mine tunnel loomed.

As fast as the steep mountainside would let her she hurried back to the canyon floor and went to find her husband. Both being exhausted they decided to rest and sleep before returning to her important discovery.

Early the next morning Mrs. Abrams led her husband up the breathtaking incline to the site. Together they investigated the cabin, finding no tools but a few rusted cooking pots. In the vicinity they saw a number of blazes on trees. On attempting to enter the mine tunnel Mr. Abrams found it collapsed a short distance inside. He and his wife felt sure they had rediscovered the Lost Cabin Mine of the Coeur d'Alenes and speculated that the original miners might have been killed in the cave-in, their bones sealed in the tunnel.

On a subsequent visit to Spokane Mr. and Mrs. Abrams boasted of having discovered "a million dollar mine" but would reveal no more than that it was located "within a day's drive of Spokane and within ten miles of a well-known and navigable bay on Lake Coeur d'Alene." The couple took back with them a *Chronicle* reporter, who wrote a news story about their discovery. He described the mine as one easy to work and located where the ore could be transported to the lake at small expense.

Yet in the end Mr. and Mrs. Abrams were disappointed. Assays of ore taken from the tunnel near the cabin proved disappointing. Theirs was not the million dollar mine after all.

So the lost mine remained a subject for discussion and speculation. A stampede which took place to Pine Creek in 1928 was caused in part by memories of the old legend. E.A. Allen who was developing a lead-and-silver vein on the west fork turned up with some specimens of gold which his neighbors found mysterious. Some said he had panned the nuggets from the creek, and hoping for another Prichard Creek strike they rushed to the west fork. When no riches of interest were found a different story began to circulate. Allen had actually bought his gold specimens from a neighbor, it was said. The neighbor had chipped them off a quartz outcropping on Latour Creek – possibly the location of the legendary mine.

Peg Allen's Lost Mine, as it is called around Medimont, is still a topic in the taverns of the lower Coeur d'Alene Valley. The mine itself is still hunted on Latour Creek. Latour is the stream next west of Pine Creek; it flows out of the south to join the Coeur d'Alene River near the Old Mission.

The Mission – it continually dodges in and out of the lost mine story. As for Father Cataldo, whose converts are supposed to have mined gold for him in the Fourth of July Canyon, his name is firmly imprinted on the scene. The Mission is now known as the Cataldo Mission and nearby lies the small hamlet of Cataldo.

The Mission church, put together more than a hundred years ago with wooden pegs, still stands on its prominent knoll, plainly seen from the modern freeway which speeds travelers up and down the valley. The oldest building in the State of Idaho, it is now a national historical monument and a shrine for visitors. Any future search for the lost and legendary gold mine of the lower Coeur d'Alene Valley might well begin with inspiration drawn from within the shadow of its imposing presence.

North Idaho

POST FALLS-HUETTER CACHE

Preston E. Onstad, a teacher of English at the North Idaho Junior College in Coeur d'Alene between 1948 and 1950, was much interested in Indian history and artifacts. He was excited to hear that a grave which might be an ancient Indian burial site had been found on the banks of the Spokane River just above Huetter, about three miles from the city of Coeur d'Alene.

Onstad hurried out to the grave site only to find an old man there ahead of him who was busy throwing dirt and bones in forty different directions – a criminally irresponsible dig which had ruined the archeological record. This outrage triggered the English teacher's temper. "I just let him have it with both verbal barrels," says Onstad.

The old fellow, in an attempt to square himself, confided in Onstad that he was looking for buried treasure. His story was that a bank in the Wallace-Kellogg area had been held up around the turn of the century, the robbers fleeing with $80,000. During the night they buried the loot somewhere between Huetter and Post Falls. The next morning they were captured and presumably hung since they never returned to unearth their cache.

The old man had not found the treasure, he admitted to Onstad, and he apologized for ruining the grave. But Onstad remained so angry he blocked out the man's name and could never remember it afterwards, though he did recall that the old fellow claimed to work at the pumping station just below Huetter.

Yet in spite of himself Onstad became somewhat interested when he heard the same tale of the bank robbers' buried loot from some old timers in a family living near the ruined grave site — and he passed the story on to me.

In trying to assess its truth I asked a knowledgeable Kellogg, Idaho, friend of mine, Hazel Corbeill, if there was ever such a batch of money as $80,000 all in one place in the Coeur d'Alene mining country around 1900. She replied that she knew of no such big bank robbery but that mining companies did have large payrolls which were in cash not checks. So a payroll heist could have yielded such a sum.

A manuscript fragment written by Tom McAndrew, who was raised at Post Falls and Lane, reveals that a bank established in Post Falls not long after the turn of the century was frequently robbed. He writes: "There were no bars on the windows and it was not much of a job for the hoodlums to knock the glass out of the rear window and go in and help themselves to whatever was inside. They worked at this job with a will." McAndrew doesn't believe the robbers got much money, yet presumably they got enough to encourage them to return more than once. In some close chase part of the Post Falls bank money may have been left behind to give rise to the story of the Post Falls-Huetter Cache.

If you go treasure hunting in the vicinity keep in mind that another cache allegedly put down by a Coeur d'Alene Indian chief, Saltice, is said to lie on the other side of the Spokane River a little upstream from Post Falls. Onstad says that a friend of his who was raised at Post Falls used to have fun during his boyhood hunting the cache but never learned for sure how it came to be laid down or how much it was supposed to contain.

In July 1968 at the Old Timers' Celebration in Harrison, Idaho, in an informal interview with Marceline Saltice Kevis, a member of the Coeur d'Alene tribe, I learned that her Saltice forebears had been war chiefs. Her grandfather Andrew Saltice, who was born in 1810, became a chief about 1840 and was a leader at the Battle of Steptoe Butte* in 1858. Her father Joseph Saltice was Andrew's youngest son by his second wife — which explains the long time-span covered by the three genera-

* A battle which is misnamed, it having taken place near present-day Rosalia rather than on the butte.

tions. The family name of these chiefs has been commemorated in the small lake of Saltice near Spokane and in a different spelling in the little town of Saltese, Montana.

Mrs. Kevis had heard no story of her grandfather Andrew Saltice having left any cache behind him but she did say he had lived at Post Falls in his later years and died there in 1902.

Since the Post Falls-Huetter cache lies anywhere within a four-mile stretch it will surely be a hard treasure to locate. Probably your best bet in seeking it would be to try to put yourself in the robbers' place and ask where in that landscape would be the most sensible spot to hide the loot.

As for Chief Saltice's alleged cache, try to locate the old Indian village and start from there. If you should discover any Indian graves, please desist and send for the nearest archeologist. He will take on the job of gathering the scientific data, leaving you free to gather the treasure.

North Idaho

OUT FROM SANTA CREEK – SOME LOST MINE TRAILS

Santa Creek is a beautiful tributary to the St. Maries River, joining it a dozen miles south and a little west of the town of St. Maries, and about two miles from the hamlet of Santa. It is great country: a clear, cool stream with fringes of alders and cottonwoods, wide, emerald-green meadows with rich grass for cattle and horses, all enclosed by mountains of sweet pine forests. In early days trout were for the taking and plenty of game roamed the hills.

Al Vanderpoel, now living in the Santa Cruz Mountains near Los Gatos, California, was raised and spent his early manhood in the Santa Creek area and has many fascinating tales to tell of the Idaho country he once knew so well and still knows through long summer visits.

He writes: "My family settled on Santa Creek in 1909, when I was small. An old Indian trail which I call the Clearwater Trail went past our place. It was the most direct route into Montana for several of the Indian tribes of Washington and Idaho, and it was over this trail that three companies of cavalry pursued a band of hostile Indians after the Battle of Steptoe Butte.

"The trail came over the Sanders divide from the Palouse country and followed Santa Creek to what is now the junction of Highway No. 95A and Highway No. 3. It left the creek there and went over the hill to Santa and followed the river to Clarkia.

"From there it took to the ridges and paralleled the Little North Fork of the Clearwater for many miles. One fork, which came out in DeBorgia, Montana, was used by trappers and prospectors. I talked to an old timer who came into the St. Maries country over this trail.

"As for the other fork, I don't know if it went on over the Bitterroot Divide or stopped when it reached the Clearwater hunting grounds. Many of the first settlers came on over the Palouse end of the trail after widening it enough to get their wagons through; and I know one old settler who came over the trail before it was widened, bringing his few possessions on pack horses.*

"For a number of years when I was small there were two annual migrations of Indians over the trail, one to hunt, the other to pick huckleberries. There would be Indians strung out for half a mile or so. A few of them had light wagons and buckboards, the rest rode horses. They would leave their rigs at Clarkia and all take to the horses. They usually camped for the night near our place and I spent as much time as possible with them. While the men put the horses out to graze, some of the women set up camp while others hung a very large iron pot over the fire. Then the women and kids would all wade into Santa Creek and catch crayfish for the pot.

"One elderly Indian woman tried to buy our dog. Mother said, 'What would you want with that fat old dog?'

---

* Charles H. Scribner of St. Maries, Idaho, retired after long years of service with the St. Joe National Forest, has written a paper setting forth the main route and branches of this trail, which he calls the Old Montana Trail. Al Vanderpoel's description of the route agrees in general with Scribner's more precise and detailed conclusions.

" 'Me eat um,' was the reply. It was no deal.

"There was one Indian couple who didn't travel with the rest but spent a good part of each summer in the Clearwater country. On several occasions when they needed supplies they offered to trade in nuggets with Charley Wells who had a ranch on Santa Creek. Charley got foxy one evening and tried to get some information by using firewater. He said afterwards it was a waste of good whisky, for the more the Indian drank the less he talked.

"A business man at Tekoa, Washington, was supposed to have the inside track with this Indian. He would trade supplies and some cash in exchange for a buckskin pouch of gold, likely at a neat profit for himself.

"According to a story making the rounds when I was a kid, the Indian woman told someone that when she was a small girl traveling with the tribe they came upon a group of whites busily panning the stream where the Indians got their gold. They sent the whites to whatever happy hunting ground whites go to and thereafter the Indians never went near the spot. She herself had never dared return but her husband had lost some of his fear of evil spirits by going to a mission school.

"A friend of mine, John Day, still living at Lewiston Orchards, a suburb of Lewiston, Idaho, confirms some of these details. When he was a teenager he was a camptender on a government survey in the Clearwater country. It so happened that the survey party put up its tents not far from the summer camp of the Indian couple. The woman was seen around the camp often, while her man seemed to be gone for days at a time. Johnny had no interest in the mine, as he knew two fellows who had disappeared while looking for it.

"Recently an old prospector came out of the Clearwater country with a story of two very old skeletons found about where I believe the Clearwater Trail Lost Indian Mine to be. He estimated the distance as a two-day saddle horse trip from where he came out on the upper St. Joe River. From his description of the country I gather he made the discovery not far from the Little North Fork of the Clearwater River.

"As in other lost mine stories there is undoubtedly some fiction in this one but there also seems to be some grains of truth. I have known John Day for many years and found him reliable.

He has offered to take me to the Indian camp site whenever I want to go. But I have had the feeling that it might be well not to disturb the ghosts who guard that tiny stream somewhere in the wilds of the Clearwater."

Al Vanderpoel is no mere armchair miner and treasure hunter. He has spent plenty of time in the hills. As he says, "I've looked for several lost mines and advise others to look for them. They probably won't find one but looking will put them in mineral country and they may find a valuable mine that isn't lost yet."

When prospecting he often uses his "papoose" rocker, as he calls it, copied from one he saw an old prospector use. It can be backpacked anywhere a man can walk, and he estimates it to be about fifty times faster and more effective than the gold pan. It is best worked by two but can be used by one. After the upper end is filled with gravel, water is dipped from the creek with a one-gallon can with handle attached and poured slowly over the gravel while the rocker is rocked with the other hand—the gravel floating out and the gold, if any, caught in one inch riffles at the bottom. Al says he thinks this kind of portable rocker has now been patented and commercialized.

Not only has Vanderpoel done a lot of prospecting but he and his partner have found more than one lost mine.

He says: "John Helsing was one of the first white men to come into the Santa Creek Valley. He was an undersized Swede with the heart of a giant, and this was the land he had come far to find.

"I grew up on a neighboring ranch and it was from Old John that I first heard the story of the Lost Mine of Charley* Creek. During the summer of 1897 two prospectors had stayed overnight at his ranch several times while they searched for a mine which they had found and lost the previous year. On that trip they had followed the Palouse River to its headwaters but had found all the good placer ground either stripped or worked. So they made their way over the mountains to a stream they later believed must have been Charley Creek. On one of its tributaries they made a rich strike: 'coarse gold from the grass roots to bedrock.'

———

* Spelled Charlie on some maps.

Of course they explored no farther, simply went to work digging the yellow metal feverishly as long as their small stock of supplies held out. Fresh meat was no problem. There was plenty of grouse, rabbits and deer at hand. But the day came when they had no salt with which to season the game.

"It was midsummer by then. They decided to make a trip out, trading some of their gold for supplies to last until winter. Making their way back over the mountains they left a well-marked trail by blazing trees at frequent intervals. It seems they somehow missed the Indian trail that went through a low pass a little to the west.

"The summer of 1896 was the summer that most of the Charley Creek timber went up in smoke. When the two miners returned with their supplies and crossed the divide they looked down on a desolation of charred snags. No blazed trail remained. They were sorry then they had not paid more attention to the direction of travel and to permanent landmarks.

"As they searched the big burn they were utterly lost. They discovered that Charley Creek is joined by two large tributary creeks before it runs into Santa Creek at Emida. Each of these large tributaries is fed by smaller creeks, which themselves have several branches. They had no idea where to look.

"They tried tracing each tributary to its source, but this was a lengthy process. The search went on until winter drove them out. When spring returned they were back again, and it was during that season's search that they sometimes stayed overnight at Helsing's ranch. The last time they talked to Old John they said they had given up and were on their way out to prospect elsewhere.

"As time went on other prospectors, including myself, searched intermittently for the lost mine. The great burn was growing up to a dense jungle of brush and saplings, and travel through it was next to impossible. The Lost Mine of Charley Creek was becoming a legend, to be added to countless other lost mine stories.

"It was my good fortune to get the rest of the story in July 1961 and it came about in a curious way. I was out with my old friend Arthur 'Hardrock' Johnson of St. Maries.

"But first let me tell you about Hardrock. He too was raised on Santa Creek and had dug his share of prospect holes. He had become a proficient trapper and woodsman. In World War II he served as a rifleman, then coming home with most of his Army pay intact he invested in a gold mine, married and fathered two children. He became skilled in log sawing, earning money to buy mining equipment. He was sure he was going to strike it rich. Then fate stepped in with a crushing blow.

"He was severely crippled by polio. The doctors thought he would do well to get his leg going enough to walk with a crutch or cane; they believed his right arm useless. But he was determined not to give in to paralysis. Every day he exercised. He forced his bad leg to hold his weight. He learned to manipulate small objects in his hand. It was a year before feeling began to return to arm and leg. He regained the use of his fingers. He took a few steps without his crutch.

"Once out-of-doors he began devising new exercises to strengthen his arm and hand. He practiced with his six-gun, using his good hand to steady the crippled one. He began operating his chain saw for a short period each day, eager to begin sawing logs again. For the family savings were long since gone and the debts were piling up. When he was able to raise his arm waist high he started to fell and saw logs, once again earning a living.

"As strength returned so did the prospecting urge. He had long heard of the Lost Wheelbarrow Mine, supposed to lie in the Hoodoo Mountains (or Thatuna Hills) in the Potlatch River drainage. This area was in driving distance of St. Maries, so he began looking over the country. While exploring a tributary to Sand Creek one day he found a place that had been extensively worked before. He knew there must be gold there but had much difficulty in tracing it to its source. One day he drove his pick into a ledge of partly decomposed quartz rock several thousand feet above the spot where his search began. When he tested the rock, he knew this was IT.

"Nearby lay some old deer antlers, so on staking his claim he named it the Buckhorn Mine. He built a cabin on the site, and later mined there several seasons. During that time he lost his limp and became able to raise his right arm above his head. Once again no man was his superior. The real gold that had come from the Buckhorn Mine was health and well-being.

"Hardrock and I had shared grub around many a campfire, both before his tough luck and after he had fought his way back to active life. I knew that where Hardrock was there was always some kind of mining excitement.

"He has invented his own mineral locator, which he keeps tinkering with. How and why the contraption works for him I will never know but I've learned when he says there's mineral nearby, it's time to become interested.

"When I looked up Hardrock in St. Maries that summer of 1961 he said he had been testing the upper Charley Creek country with his mineral locator and had picked up a gold deposit but hadn't had time to check it out. I suggested we make a trip up there and see what we could find.

"The following morning we bounced along in Hardrock's four-wheel drive for miles over an apology of a road, then parked the jeep just off into the brush to give others a chance to get by. We donned our packsacks, and as soon as Hardrock got cross angles on his mineral locator to give us the direction of travel we set off.

"We pawed our way through brush for some distance and then entered old growth virgin timber where the going was much easier. We soon came out on a tiny stream.

"Ahead of us were the remains of a dam and of a sluiceway constructed of split cedar planking. Partly overgrown with brush was a neatly piled row of rocks, the unmistakable mark of the old time placer miner.

"The pieces of the puzzle fell into place. It was the Lost Mine of Charley Creek! We had hit it square on the nose. We couldn't have walked straighter to it if we had known its location and lined it up with a compass.

"We both had a feeling of smug satisfaction at finding it so easily. But our smugness received a severe jolt a few seconds later when we heard voices on the wooded slope above us. We investigated and found a family already on the site, already mining. The Lost Mine had been found ahead of us.

"The reason the two original prospectors had failed to locate it again was that it was not in the burn at all but just outside in the virgin timber. Everybody had been looking in the wrong place for it."

But this denouement failed to discourage Al and Hardrock. They kept up their summer investigations. While looking for a lost mine in the Hayden Lake area (the Lost Breen Diggings?) they discovered a good gold prospect. While searching for a lost mine on the North Fork* of the Coeur d'Alene River Hardrock found a fine silver prospect—one he and Al have opened up with such success they have received an excellent offer, not yet accepted, from a big silver mining company.

Last heard from Al Vanderpoel was en route from Los Gatos to North Idaho to take care of his various mining interests. "Silver is the big deal now," he says. "It may turn out to be a bigger boom than the old gold rush. It ought to be good for years."

Certainly he has proved the soundness of his own advice: Get out into mineral country. You may track down that old lost mine. Or while looking for it you may find a rich mine that's not lost yet.

---

* Now officially designated the upper Coeur d'Alene River.

North Idaho

THE SPOTTED LOUIE MINE

Those who knew the Indian who gave his name to this lost mine say his skin was noticeably spotted, apparently from some skin ailment. So he was always called Spotted Louie and his full name has been lost to history.

As I first heard the story, he used to spend time in Bovill and Elk River, Idaho, paying his way with gold. When his poke was empty he would head east into the Clearwater River country, be gone for several months and return heavy with a new supply of gold. Old timers said he must have found a rich lode mine because his gold appeared to be hand ground out of quartz. Definitely it was not water worn as placer nuggets would be. But Spotted Louie was very close-mouthed and no one was able to figure out where his mine lay.

In the 1920s a man I'll call Bill Smith (he expresses an aversion to being identified in print) spent some months in Kooskia, Idaho, and became acquainted with an Indian woman, Martha Yellowbear, who had pitched her tepee just north of town. Both she and her daughter were excellent shots and Bill was amused to discover that if any of the town's canine pets strayed out that way Martha's tepee was soon afterwards redolent of dog stew. Martha told Bill that the tastiest dish she had ever cooked had featured the pampered pet of the town banker. She was impatiently waiting for another fat dog belonging to a certain merchant to come strolling by.

Spotted Louie was dead by then but Bill had heard the talk about his mine and was intrigued to discover that Martha Yellowbear had been a friend of his. She liked to talk about Louie and brag what a good mine he had and how none of the whites were ever smart enough to track him back to it. This encounter convinced Bill that the Spotted Louie Mine was real enough, but he could get no helpful clues out of Martha as to its location. He doubted that she ever knew where the mine was or cared. It had been enough for her to help Louie spend his gold whenever he came to town.

It was Bill Smith who pointed out to me on a map of the St. Joe National Forest a stream named Spotted Louie Creek. With headwaters near Bathtub Mountain it runs southwest into the Little North Fork of the Clearwater River near an eminence called Spotted Louie Point. On most modern maps the name is spelled Louis, which sparks Bill's wrath: "Evidently some collegiate intellectual decided that Spotted Louie wasn't refined enough and so had to change it."

But were the creek and point actually named for the Indian of the lost mine legend, and what is the proper spelling? I put these questions to various people but found only one man with any knowledge on the subject. He was Jim Estes, who had done twenty years of duty in the Roundtop District of the St. Joe National Forest, part of the time as fire control officer.

In reply to my query Estes wrote: "In the middle fifties an Indian from the Coeur d'Alene Indian Reservation at DeSmet, Idaho, worked for me at Roundtop. He was a young man, I've forgotten his name, and he's dead now anyhow. I asked him if he knew the source of the name Spotted Louis. He told me it was named after an Indian known as Spotted *Louie,* and said he heard his grandfather and other old Indians talk about trips made from the prairie country (around Tensed, DeSmet and Plummer) into the Spotted Louis Creek area to bring out *gold.*

"I have been pretty well all over the head of Spotted Louis Creek chasing smoke many different times during the last twenty years but never saw anything resembling a mine. I have also fished the creek but noticed no diggings."

Estes went on to say he believed that Louis or Louie was a surname among the Coeur d'Alene Indians.

As this lost mine story had seemingly originated in the Bovill-Elk River area (at least the version I had heard), the idea that the Indian in question could be of the Coeur d'Alene tribe came as a surprise. So did the suggestion that Louis or Louie might be a last name instead of a given name.

On a trip to North Idaho in the summer of 1968 I was able to substantiate the fact that indeed Louie, not Louis, is a well-known surname among the Coeur d'Alene tribe. In an interview at the Old Timers' Celebration at Harrison, Marceline Saltice Kevis, a member of the Coeur d'Alene tribe, and her husband Charles Kevis, assured me that Spotted Louie, now dead, had been a Coeur d'Alene Indian. The place on which they lived at Worley once belonged to him. They did not know his proper first name and had not heard the story of his mine.

In St. Maries I called on Ray L. Hilding, then Supervisor of the St. Joe National Forest, to discuss the correct spelling of the creek name. Looking back on old Forest Service maps he and Forest Technician Lawrence Glover discovered that indeed the name had appeared as Louie until about 1940, when it had undergone the change to Louis. They agreed that the Louie spelling is correct and promised that the name would be shown correctly on future maps of the St. Joe National Forest.

So it seems amply established that Spotted Louie was a Coeur d'Alene Indian who left his name on a tributary to the Little North Fork of the Clearwater River, and if there is any truth in what the young Indian told Jim Estes, the Spotted Louie Mine should lie somewhere in the vicinity of the creek, since members of the Coeur d'Alene tribe used to travel to the area to bring out gold.

Along the Monument Trail

And yet lost mine tales seem born to confusion and contradiction. There is a completely different story called the Lost Louis Sam Mine, which may well pertain to the Spotted Louie Mine. The Indian in this case was also of the Coeur d'Alene tribe and the similarity of name causes one to wonder if correctly this mine should not be called the Lost Sam Louie.

This Indian too went out from the DeSmet Reservation each spring traveling east into the mountains. He followed a well-defined trail that led along the top of the St. Joe-Coeur d'Alene divide. According to Orland A. Scott in his book *Pioneer Days on the Shadowy St. Joe,* this is the Monument Trail (being named from the number of rock cairns or monuments which mark its course) and is "the historic Indian route into Montana."

The Indian was crossing the headwaters of one of the streams flowing down from the divide toward the St. Joe River when he came upon a waterfall and in the pool below it saw the golden gleam of nuggets. Each year thereafter he would return to the stream to harvest the new pieces of gold brought down into the pool by the spring freshet.

Unfortunately no single clue in this story tells us how far along the Monument Trail the Indian had traveled before he made his lucky discovery. So the stream could be any one of a dozen creeks all the way from Hell's Gulch near St. Maries to some wild canyon east of Avery. Such an area takes in all outdoors along the northern watershed of the St. Joe Valley and is too wide of the mark to be of much interest to treasure hunters.

And, contradiction within contradiction, still another story exists as to how and where an Indian, this time called Lewie Sam, made a gold discovery. In this version he and a partner were out deer hunting in the upper St. Joe Valley not long after the 1910 fire. They shot a deer which fell into a steep ravine and on descending into the declivity to reclaim their venison they discovered a gold vein along the face of the cliff.

The partners swore each other to secrecy, with a threat of death to the one who first revealed the location of their strike.

But talk of the Indian discovery got out and white men eagerly hunted the mine in the upper reaches of the St. Joe Valley. And the search continued over such a long stretch of time that as late as 1960 one man claimed to have found the lost Indian bonanza. George Horine of Spokane, Washington, said that he and his son, Albert Horine, had located the mine on the upper end of Jessie Creek, twenty miles northeast of Avery. They had sent specimens from the vein to Portland, Oregon, for assaying and had received a report which promised 128.64 ounces of gold to the ton, giving a ton value of $4502. Horine made news when he appeared in Wallace, Idaho, county seat of Shoshone County, to file location notices blanketing the site of the discovery and announcing that he and his son would name the property the Lewie Sam.

So does legend lead and mislead. Louie, Louis, Lewie – all pronounced the same – could have been one wide-ranging member of the Coeur d'Alene Indian tribe. Or Louie, Louis, Lewie may have been three different Coeur d'Alene Indians, each with a notable mining adventure. With the Spotted Louie Mine allegedly lying in the drainage of the Little North Fork of the Clearwater River; with the Lost Louis Sam Mine a sparkling pool of nuggets not far off the Monument Trail at the headwaters of a northern branch of the St. Joe River; and with the Lewie Sam Mine lost and found again on Jessie Creek – which way lies the truth?

AN OLD GREEN COPPER STILL

This concerns an entirely different kind of lost treasure – and a rare variety, at that.

In November 1966 Bennie Toland of St. Maries, Idaho, was deer hunting in the upper east Bond Creek region of the St. Joe River drainage when about a half-mile northwest of the abandoned Gold Ridge Lookout Station he happened on an old wooden barrel, the kind moonshiners liked to keep mash in.

The place was certainly out of the way enough to make a good hide-out. Seven or eight miles away Bond Creek joins the river at the old ghost town of St. Joe. Even in the heyday of the mill town a still on upper Bond Creek would have been located a long distance from potential customers; but it is true that the economics of moonshining was such that the finished product could stand a lot of transportation costs.

What made Toland's find doubly interesting was that a friend of his, Lee Carpenter of St. Maries, had talked to a hunter from Coeur d'Alene who said that while he was going through that same area the previous fall he had come upon two copper boilers with copper coils attached, still in excellent shape and obviously a part of a moonshining outfit.

This report seemed to indicate that there had been recent moonshiners in the Gold Ridge area. About a week after the discovery of the mash barrel, Bennie and Lee's son, Little Lee (so called even though he is a big man) took time out from a hunting trip through that part of the woods to search for the moonshiners' outfit. They did find a much older copper boiler, green with age, and all smashed in except for the dome, which was in fair condition.

Nearby were pieces of broken jars, parts of an old stove, a granite pan and granite pitcher, and some rusted bedsprings – all evidence that a cabin had once stood on the spot.

Lee Carpenter, Jr., is still hoping to find the two newer boilers and pack them out of the hills; and no doubt the discovery of a bona fide moonshining outfit would be as satisfying as dig-

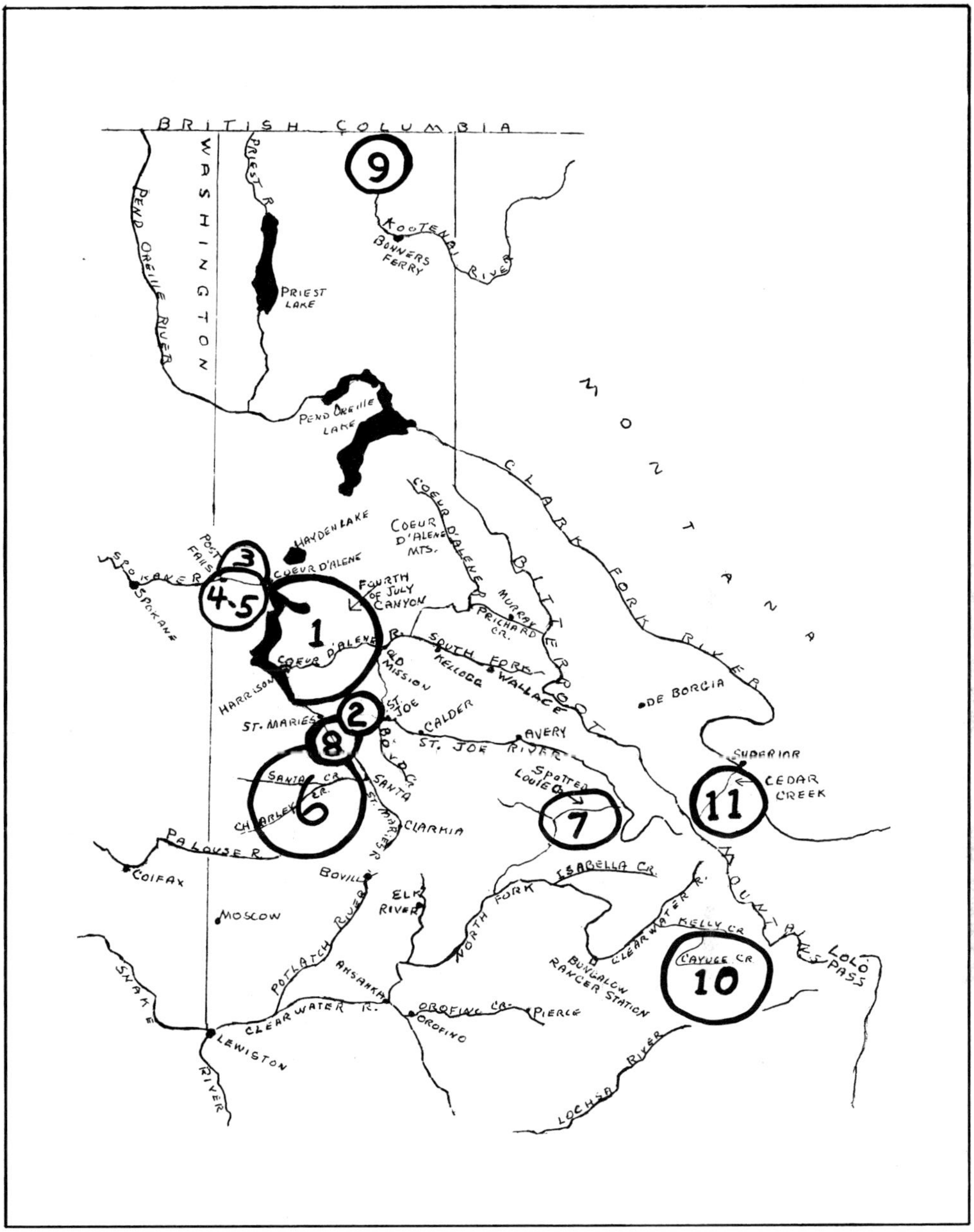

IDAHO PANHANDLE AND CEDAR CREEK, MONTANA

1—Legendary Lost Gold Mine of the Coeur d'Alenes
2—Lost St. Joe Gold Mine
3—Post Falls—Huetter Cache
4—Chief Saltice Cache
5—Post Falls Diamond Mine
6—Santa Creek lost mine tales
7—Spotted Louie Mine
8—Lost Spanish Silver Mine
9—Ancient tamarack tree blaze
10—Indian Post Office Mine
11—Chinee Cache of China Gulch

Mystery of Chinese coins found among early-day Clatsop Indians has been revived by a curious Chinese coin—the two faces of which are pictured here—picked up from a beach near Tillamook, Oregon, by Dan Lawrence in the 1950's. (Mrs. Loran Lawrence)

ging up a buried treasure chest, and would create about as much interest.

If you think that search for the hidden still, made in a cold afternoon rain, distracted those St. Joe boys from their hunting you are mistaken. Each bagged his deer before starting the long trek back down Bond Creek.

Extreme Western Montana

CHINEE CACHE OF CHINA GULCH

In 1868 a gold rush took place to the St. Joe River valley which is recorded in history mainly because the excitement lured a French-Canadian miner, Louis A. Barrette, from California. When he failed to find paying ground on the St. Joe he continued on east and crossed the Bitterroot Divide. From one vista on the trail near the summit he could look down into the canyon of Cedar Creek — a stream which joins the Clark Fork near the present town of Superior, Montana.

A basin he saw appeared promising as a place to prospect for placer gold. But being too short of provisions to remain he went on to Frenchtown, where he not only spent the winter but loitered through the next summer. It was the fall of 1869 before Barrette and a friend returned to Cedar Creek.

On Cayuse Creek, one of the little tributaries to the Cedar, they struck rich gold. The news soon leaked out and a midwinter rush took place, so that by the spring of 1870 nearly all of Cedar Creek, its tributaries and gulches were staked. Several mining towns sprang up, chief of which was Louisville, named for Barrette; Forest City, four miles away; and Mayville, at the head of the creek. Three years of hectic mining activity ensued.

Chinese came into these camps very early. It is told that in 1871 unscrupulous white miners bilked some Chinese miners of $30,000 by selling them a "salted" gold mine. And in 1872 when a Forest City store reported the disappearance of a small iron safe containing several thousand dollars in gold dust, the Chinese were naturally blamed – *naturally* since white miners found the Orientals such convenient scapegoats. A vigilance committee abused some Chinese at a local washhouse and also shot and killed a party of Chinese travelers. The safe with its gold dust was recovered, though it is by no means clear that the murdered Chinese travelers were guilty of the theft.

In 1873 the Cedar Creek boom collapsed and the white men stampeded out of the canyon as rashly as they had stampeded in. This left the Chinese in possession, and as so often happened in the west, they stayed on in the abandoned camps to rework the diggings a second time. The Chinese were patient and diligent and as they lived economically they needed no such spectacular cleanups as the Americans demanded in order to deem the mining worthwhile.

Over the years the Chinese miners of Cedar Creek accumulated small fortunes by slow and careful digging, and one by one they left to return to China or San Francisco, or to go wherever fate took them. At length the only man to remain was one who had been ostracized by his fellow miners. Just what trouble had made him a pariah among his own people the white folks at Superior, the nearest town, never bothered to learn. Perhaps he had cut off some one's queue, violated a sacred taboo, or stole some other Chinaman's money. Whatever it was, he was ordered to remain and do eternal penance in the isolated canyon – on penalty of death if his countrymen should ever find him elsewhere.

So he stayed on for twenty lonely years, living in a hovel in a gulch off Cayuse Creek known as China Gulch. He continued to work the already-worked placers and bought his meager supplies at Hurd's store in Superior, paying in gold dust. Since his wants were few and his expenses almost nil and yet he was known to be an industrious man, the people with whom he dealt and who knew his story became convinced that he was laying by part of the gold dust he was almost certainly digging. So by the time the Chinese miner died around 1900 a ready-made story existed about the cache he must have left behind, secreted somewhere in or about his cabin.

Immediately after his death a large traffic sprang up between Superior and China Gulch on Cayuse Creek as people trooped in to look for the treasure. The hovel where the Chinaman had lived was torn apart and many holes put down in its immediate vicinity.

When none of the treasure hunters met with early success, talk of the cache spread farther afield and reached the ears of Clayton McBride. He was a trapper living in a cabin on Graves Creek, with a trapline strung out all the way up to Chamberlain Meadows on the St. Joe-Clearwater Divide. He had earned a certain fame as an artist in the making of sourdough hotcakes. But trapping was a hard and lonely business, times were tough, and McBride decided he might make easier money by migrating to Cedar Creek and picking off the Chinee Cache.

His idea was that people had been conducting the search too near the site of the cabin. After all, a Chinaman smart enough to hoard the gold in the first place would be smart enough to know that the first place snoopy people would look for his hard-earned fortune would be in and around his cabin.So it was likely he had buried it a fair distance away, and Clayt felt that if he could visit the gulch and stand on the ground he could, by a process of deduction and elimination, figure out where the cache lay.

So Clayt quit his Graves Creek cabin and moved to Cedar Creek and there became a dedicated treasure hunter.

But a friend of mine who knew Clayt McBride says he is sure Clayt did not find the gold dust cache. For if he had made his fortune he was the type who gladly would have become a man of leisure. Instead he spent his last years running a small restaurant in Superior – specializing, it is hoped, in those delicious sourdough hotcakes. He died in Superior about 1940.

In later years there was other mining activity along Cedar Creek. At one time the canyon was turned upside down by a dredge. At the head of the creek the dredge was left to rust near the old site of Mayville. Between 1900 and 1907 the Amador copper mine was in operation, and more recently the Gildersleeve mines.

Jean Davis, compiler of *Shallow Diggin's*, tells that one of these later mining operators became interested in the Chinese relics he found among the old Cedar Creek placer diggings and took to collecting them. Unless he collected the cache left behind by that most unfortunate soul, the last Chinese miner of Cedar Creek, it may be there yet under some rock pile in China Gulch.

South Central Montana

MINER'S LOST BOOTFUL OF GOLD

Here is a strange tale which for long years has intrigued Montana treasure seekers and puzzle solvers. As an enigma it should admit of solution since it concerns a definite marker with recognizable signs and numbers.

But to date the puzzle has stumped all comers — a double entendre since the conundrum is found carved on a six-foot stump a couple of miles east of Cooke City, Montana. Happily, there is a buried treasure story which explains why the hieroglyphics were left on the spot.

The carving looks something like this:

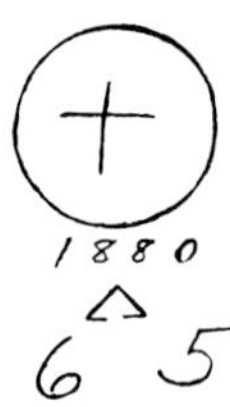

The circle is about eight inches in diameter; the cross inside is about three inches. The numbers 1880 are of medium size. Some believe the partial triangle just below is really the outline of a tepee. The numbers 6 and 5 are large and dominant.

Cooke City, locale of the mystery, is not a city at all, only a mining hamlet up in the mountains on the northeast corner of Yellowstone Park, not far from the Wyoming border. Jay

Cooke, Jr., son of the financier, was once interested in the area and the town was named for him, but the collapse of his father's financial empire, J. Cooke and Company, was a disaster which put the Northern Pacific Railroad into receivership and brought on the terrible panic of 1893. So the name was not a lucky one and Cooke City's dreams of becoming a metropolis all failed: The smelter didn't pay, the railroad didn't arrive, and the mines didn't prosper.

Soon the old timers of Cooke City had nothing to do but move from sun to shade, dreaming of ways to decipher the markings on the stump not far from town and locate the buried gold to which it purportedly is the key.

According to belief, the figure 1880 is the year the cache was laid down. It was then that two miners who had accumulated a fortune from the gulches surrounding Virginia City, Montana, decided to return east. Understandably they were concerned about safeguarding the hoard which would make them wealthy men in their old homes. They feared the well-traveled roads where robberies were all too frequent and where even treasure shipments were sometimes hijacked. They decided the best thing to do was to put the gold in packs, load it on a mule and lead the mule through the wilderness along paths so solitary no outlaws would be waiting there.

Unfortunately for their carefully laid plans, certain ruffians in Virginia City who knew they had hit it rich learned of their projected trip. So when the miners made what they thought was a quiet unobserved exit from town, they were secretly followed.

The two miners followed an old Indian trail which led up the Madison River and across Yellowstone Park (then a virtually unknown wonderland). They then made for the Clark's Fork of the Yellowstone (a stream not to be confused with the Clark Fork of the Columbia River). The Yellowstone would take them to the Missouri River and on to civilization.

In the vicinity of the present Cooke City (where only a mine or two existed at the time), they camped some yards off the trail. Taking the packbags of gold off the mule they secreted them behind a log, then turned the animal loose. Suddenly shots rang out. The grazing mule fell dead. So did one of the miners. The other miner managed to throw himself into the brush and to escape in the darkness. The attackers ramsacked the camp but did not find the gold. When the surviving miner was sure they had departed he crawled out from his hiding place.

With no pack animal he could not, of course, transport all the gold. He secreted what he could about his person, and then, having no other receptacle, he removed one of his dead partner's boots, filled the boot with gold and buried it. He carved the markings on a nearby tree as an aid in locating the cache when he returned, and he also drew a map of the site before he continued on down the Indian trail toward the Yellowstone River.

Cooke City had its beginnings not long after this tragic event when the whitened bones of the mule could still be seen not far from the marked stump. But the significance of the bones and the hieroglyphics was not known for some years. Then a nephew of the surviving miner, together with two friends (or, in some versions, three nephews), arrived in Cody, Wyoming, and engaged Sam Launchberry to guide them fifty-five miles northwest into the Cooke City area. The members of the search party had in their possession the map the miner had made when he buried the gold, and with Launchberry's expert aid they located the stump with its cryptic signs. But the nephew (or nephews) had not sufficiently mastered, or perhaps had even misunderstood, the meaning locked up in the symbols and the party failed to find the gold-filled boot.

In the usual interpretation, the figures 6 and 5 indicate that the cache is sixty-five feet — or yards — from the stump. But the job of digging up the whole perimeter of such an outsize circle has proved too tremendous and it has never been done. Besides, the figures could just as well mean six feet in one direction and then five in another.

Some years ago the property on which the stump stood was slated for development as a summer resort but it is hoped that the developers spared the stump, realizing that it would be a big drawing card as the physical evidence of an enigma, which, if solved, would reveal the burial place of that Miner's Bootful of Gold.

PART TWO
TREASURE TALES TWICE TOLD

Old lost mine and treasure stories never die. They do not even fade away. If they have the real stuff of life in them they only come on stronger with the passage of years.

The pieces in this section are "twice told" because they involve —

New searches

New clues

Readers' comments

Questions and Answers

— which elucidate, give luster to, or even make important corrections in the record of the stories I recounted in my former book, Lost Mines and Treasures of the Pacific Northwest *(Portland, Oregon: Binfords & Mort).*

For the convenience of readers I have provided summaries of the previous stories so that the added material may be understood and enjoyed by those who do not have the other book at hand. For those who wish to read the stories as originally told I have provided page numbers for easy reference.

TREASURE CHESTS AND BEESWAX

THE MANY MYSTERIES AT NEAHKAHNIE; PIRATE CHEST OF THREE ROCKS BEACH (Hult, *Lost Mines*, pages 1-45).

No other Pacific Northwest treasure tale contains more separate elements of mystery than the one embracing the Oregon coast from the Columbia River south to Neahkahnie Mountain and Nehalem Bay. Earliest explorers, traders and settlers were confronted by a number of riddles: a red-haired Clatsop Indian where no red-haired natives should be found; chunks of beeswax, some with cabalistic markings, washed up from Nehalem Spit; Chinese coins ornamenting Indian garments; Indian tales of shipwrecks, two or three; the ribs of old ships uncovered and covered by shifting sands; and a legend that a crew from a winged ship carried a chest up Neahkahnie Mountain and buried it there.

As early as 1848 an Astoria pioneer, John Hobson, on a visit to the Nehalem River mouth, heard directly from the Indians that "long before they were born, the wax vessel was lost on the spit, and another anchored near the shore, and some people brought a chest up on Necanny Mountain and carried sacks of money and put them in a chest and killed a man, and put him also in the chest. Afterwards they marked a stone, or very large rock, rolled it on the chest, and went back to the ship and sailed away."

But Hobson was unable to locate the marked stone and so dismissed the story as unreliable. Others were less skeptical, and shortly after 1865 Hiram Smith, a Tillamook pioneer, began digging in an attempt to locate the chest. His son Pat Smith became one of the most dedicated of Neahkahnie treasure hunters, devoting his whole life to the cause. Around 1890 Pat Smith discovered on the mountainside a large glyphic rock which he moved to confound other treasure seekers. He also found two or more smaller marked rocks, one of which, showing dots on its surface, he considered the key to the mystery. But Pat Smith died disappointed.

Before that time he had taken a partner, Charlie Pike, to whom he passed on the lore he had gathered. Pike in his turn told the treasure secrets to a later partner, Milo Merrill of Bremerton, Washington. Merrill became convinced that the prize was but one of twenty treasure chests cached by the Spanish government at prospective sites of settlement in the New World. In his search Merrill used a bulldozer and drilling equipment.

One school of thought identified the Nehalem wax as mineral wax indicative of a petroleum deposit. This started on oil boom. Actually the wax has been established as Oriental beeswax by chemical test and reasons for believing it to be part of the cargo of a wrecked Spanish galleon are convincingly presented by William Lytle Schurz in his book *The Manila Galleon.*

At Three Rocks Beach about fifty miles south of Neahkahnie the skeleton of a man eight feet tall identified as that of a Negro was unearthed in 1931. A local Indian legend tells of a ship blown into the mouth of the Salmon River and wrecked, the crew carrying a huge box or chest ashore and burying it. The crew members then departed from the site and walked overland, all except two left behind to guard the treasure. One of the two was a black man.

More than a century of research and digging have failed to solve any of the enigmas that haunt the Oregon coast. If anything the mysterious aura has grown thicker and darker over the years.

The Five Teakwood Boxes

It is certainly foolhardy to bring forth at this late date a new, different and even contradictory Neahkahnie story which tells of not one but *five* treasure chests buried in a stream bed cache just south of the Nehalem River. I myself regard the story with some skepticism but whenever I have recounted it to Neahkahnie buffs they have reacted favorably. As one experienced Oregon coast treasure hunter says, "I think the story has merit because too many things told about the Neahkahnie Treasure have never added up." So I pass it on for whatever it is worth.

The story was related to me in 1958 by F. L. Watrous, a treasure hunter whom I had mentioned in connection with his search for the Lost Tillamook Mine. He wrote to say he had come across another tale he believed I would find of interest. As I was living in Portland, Oregon, at the time and he only thirty miles away at Forest Grove I made the short trip necessary to interview him in his home.

Watrous said that when he had first begun to pursue information about the Lost Tillamook Mine he had visited the Siletz Indian Reservation and had questioned the natives there. He had talked to a well-dressed, apparently well-educated, elderly Indian who lived next door to Hall's reservation store. This Indian said to Watrous:

"I'm afraid you won't find out much about the Lost Indian Mine around here. We can't tell you because we don't know anything. The only story along that line I know of is one in my own family. Maybe you've heard something about it. My grandfather was a Portuguese sailor and helped bury some boxes of Portuguese coin down on the coast. He went back to look for them and so did my father but they could never find them."

Portugal in his grandfather's time, said the old Indian, was an extremely poverty-stricken nation and many people wished to emigrate to the new world. One group sent emissaries who made an agreement with Dr. McLoughlin of Fort Vancouver, the Oregon headquarters of the Hudson's Bay Company, to set up a colony on the Santiam River. But it was some three years before all arrangements could be made and the colonists sailed in two ships. When the ships arrived at the Columbia River bar it was learned that Great Britain and the United States had effected a boundary settlement. The United States had received title to all the land north as far as the 49th Parallel and the Hudson's Bay Company was no longer in control of Oregon.

In the general disappointment and confusion most of the colonists decided to turn back and were transferred to one ship. The second ship, being laden with merchandise meant for opening two colony stores, remained behind to try to dispose of the goods in trade with Indians along the coast and in the Columbia River. This ship also carried the treasury of the colonizing group in Portuguese coin stored in five teakwood boxes.

No sooner than the captain had moved his ship down the coast to begin trading than he began to feel the heavy responsibility of having charge of the colonists' money. He worried over losing it should the ship be attacked by Indians at some trading stop. The safest course would be to cache it on shore. He would then trade on up the coast and into the river and stop by to pick up the money when he returned this way heading for home.

He ordered the teakwood chests loaded into a longboat, selected a mate and crew to man the boat and gave minute instructions for making a cache which would remain undetected. On reaching shore members of the work detail tied the longboat to some driftwood logs and walked inland a little distance, where they diverted the waters of a small creek. Excavating a hole in the stream bed they lowered the teakwood boxes into the hiding place, covered them over well and then put the stream back in its former course. All traces of the digging were cleverly obliterated.

Even as they worked a sudden, fierce storm enveloped them and when they returned to the shore they discovered the longboat gone, whipped from the beach by the squall. They were stranded and in the thick eye of the storm could not sight their ship to hail it. But surely their captain would soon realize their plight and come to rescue them.

They huddled along the shore that night and in the morning saw the ship to the south of them seemingly in trouble dragging her anchor and drifting toward the surf on the incoming tide. Just as it seemed the vessel might come close enough for them to board the tide changed and she drifted away again.

The longboat crew had brought little to shore with them, only some cutlasses and one blunderbuss with very little ammunition. Hungry and cold they built a driftwood fire and managed to catch some crabs. They spent another night on the beach and in the morning were dumbfounded to see that the ship had struck bow on beyond the mouth of a river a little to the north of them. They could see Indians looting her.

Alarmed for the lives of their captain and shipmates they hastily put together a raft of driftwood cedar roots and crossed the river to try to effect a rescue. But the Indians came away from the wreck and met them menacingly with guns which apparently lacked ammunition, since in the skirmish which followed the natives fought with bows and arrows. One of the longboat crew was wounded and the white sailors were forced to retreat across the river.

They remained with the wounded man until he died and then decided to start overland for Fort Vancouver, which they thought to be about fifty miles distant. Afraid of the Indians along the coast they struck east into the mountains. They found the going unbelievably rugged with thickets of tropic-like vegetation including devil's club, salal and salmonberry, and except for a trout or two they caught and some rabbits they snared they had nothing to eat. It was fall by now and the storms were severe. As a result of the privation, cold and hunger one by one the men died. The sole survivor came out on the east side of the mountains and was discovered in a delirious condition along the trail by Iler, the homesteader on Iler Creek, who was at the time out cutting wood.

The sailor recuperated that winter on the Iler ranch, helping about the place and even telling his benefactor about the teakwood chests of coin he had helped bury on the Oregon coast. This exotic story got around among the neighboring homesteaders, one of whom was White of White's ranch.

In the spring Iler and eight of his neighbors persuaded the "Spanish sailor," as they called him even though he was Portuguese, to guide them over the mountains to the site of the cache. En route, however, he began to regret his decision realizing that by the time the ten of them divided the spoils he would gain very little — too little to repay him for the terrible experience he had been through. On reaching the coast he pretended not to be able to find the right stream bed.

The sailor remained with the Ilers until early summer. One of the local girls had become attached to him and he even considered marrying her. But he had left a sweetheart in Portugal and was becoming more and more homesick for his native land. He set out for Fort Vancouver where he found a ship sailing for Spain. He made his way home and there married his first love. But he could not forget the teakwood boxes of coin and was determined to return to America and retrieve the fortune. Traveling to London he apprenticed himself to the Hudson's Bay Company for fifteen years. So arriving again on the Columbia River he became a clerk in the Oregon headquarters of the Hudson's Bay Company and was in due time able to send for his wife and the son who had been born to them.

Before his years of apprenticeship were up he was nearly blinded by a powder explosion and was released from his employment. He then moved his family to the Oregon coast and began a careful search for the stream where he and the other sailors had made the cache. But the ravages of time had so changed the terrain he had no luck in his quest.

The sailor's son who was about ten years old when the family moved to the coast became interested in the search and continued it even after his father became too old. The son married a Nehalem Indian woman and to this union was born the half-breed Indian from whom Watrous heard the story on the Siletz Reservation.

Watrous believed that the beach where the longboat landed and near where the cache was made was slightly north of the present location of Manhattan, Oregon, and the Portuguese ship was wrecked north across the mouth of the Nehalem River in the vicinity of Neahkahnie Mountain. The problem is somewhat complicated by the fact that the Nehalem River mouth has shifted over the years.

The story seems to proceed logically and convincingly. It is quite detailed and displays a good knowledge of the Oregon country. If it contains no historic truth, who could have made it up? Watrous was about sixty-five years old when I talked to him, not so much the grizzled prospector type as just an ordinary workman living a bachelor existence in a modest little house. Surely he was too prosaic to have invented the intricacies of the tale. And the half-breed on the reservation seems unlikely to have been its creator. Could one of the two have read it somewhere? Again unlikely since if it were a published story surely it would be known among Neahkahnie Treasure fans.

As for historic corroboration the very fidelity to detail which marks the story tends to be its undoing. The arrangements for the Portuguese colony are said to have been made just before the boundary settlement of 1846 – quite possibly in 1845. Since the ships came three years later the main events apparently took place around 1848. This checks out since S.W. Iler crossed the plains in the summer of 1847 and in the fall claimed homestead land near what is now Gales City. The creek on which he settled took his name, though unfortunately it appears on some maps today as Iller or even Miller Creek. Efforts are being made to restore the correct spelling of Iler. The White family took a square mile of homestead land along the second stream north from Iler Creek, giving their name to White Creek.

The year 1848 is fairly late in the history of the northwest coast and is well documented. If any such Portuguese colony was projected for the Oregon country the facts should be known and if a Portuguese ship was wrecked near Neahkahnie or elsewhere in that vicinity marine histories should carry some mention of it. As yet no confirmation of any of these details has come to light.

Moreover the basic logic of a number of the events recited is questionable. (1) Even though the Hudson's Bay Company's sphere of influence had been pushed north of the 49th Parallel there is no reason why the Portuguese emigrants could not have settled on the Santiam anyway. The United States wished to fill up the vast land of Oregon and would have welcomed the colonists. (2) Why should the colony's funds have been left aboard the second ship when it was the emigrants who would need money wherever they went whereas the captain would be trading and have little use for coin? (3) How could the captain have felt it safer to bury the boxes of money on a wild, unknown shore, inhabited by Indians, than to have kept them under his personal supervision and defense?

And in point of fact Fort Vancouver was no longer a Hudson's Bay outpost after 1849, so the Portuguese seaman could not have returned to the Columbia River in connection with his duties as a Hudson's Bay clerk. By then Fort Victoria on Vancouver Island was the western headquarters of the fur-trading company. It seems unlikely that so traditionally a British company would have hired a Portuguese national, and as for a fifteen-year apprenticeship, no other reference to such a long term contract has been found.

Yet to debate the other side of the question it is true that not all events get recorded in history, or obscure records are ignored by disinterested historians. Not all human decisions are sensible. A fairly high percentage of them are riddled with error and contradiction. And it is true that a crew of shipwrecked sailors hitting east across the coast range from around the mouth of the Nehalem River could reasonably have been expected to follow an old Indian trail which would have brought the lone survivor out in the vicinity of Iler Creek.

All in all the story is a puzzling one. Possibly it is a tall tale trying to mask itself as truth, except in that case it ought to stick to the facts of conventional history. Possibly it is sheer invention, except that the manner of telling does not smack of the craft of the fiction writer. In plot it is as accidental and unpredictable as life itself. Perhaps the facts as presented have been distorted by many retellings and the bones of the story are those of some forgotten but tragic shipwreck.

The Beeswax Dated

Thanks to the initiative of Burford Wilkerson, former manager of the Tillamook County Chamber of Commerce, the beeswax from Nehalem Spit has been dated by modern radiocarbon dating methods.

Wilkerson supplied to scientists of the Shell Development Company a piece of beeswax from a block which is kept on display at the Tillamook County Pioneer Museum and which is carved very plainly with the number "67" in large size. In the photo reproduced in this book the famed block is held by Alex Walker, curator of the museum.

In its December 1961 issue *Shell News* reported the results of the Shell Development Company tests. The first test used a modern piece of wood as a standard of comparison, since modern beeswax has been exposed to so much radioactivity from nuclear explosions as to make it untrustworthy for this purpose. Later on Kenton Knight of The Dalles, Oregon, supplied a piece of beeswax formed in 1937 which had been protected from radioactivity. New tests were then made using both the

Knight beeswax and the modern wood sample. Careful calibration indicated that the Nehalem Spit beeswax was formed in 1680, plus or minus 110 years. This range of time would take in the 1670s, the decade some treasure fans believe is represented by the marked beeswax block, and also the 1705 disappearance of the Spanish galleon, *San Francisco Xavier,* which may have come to grief on the Nehalem Spit.

Regarding an astounding item which came out of one block of beeswax I have an interesting letter from Mrs. Norman P. (Betty) Allen of Portland, and it is provocative to think that if her find had only survived to be examined by scientists and historians it might have helped to date the beeswax much closer.

She writes: "When I was growing up I spent all my summers with my parents, Mr. and Mrs. Charles A. Fowler, Jr., at Twin Rocks, south of Neahkahnie Mountain below Rockaway. We were fascinated by the legends concerning the treasure. Many of these were told us around the fire by Mrs. Helen S. Watts, longtime resident of the area.

"When I was in the seventh grade in 1930 I found a chunk of beeswax buried in the sand. It was about five or six inches in girth and irregular in shape. On examining it we noticed a piece which seemed loose. My father pried it open and we discovered inside the corners of some pages which seemed to be a part of a religious book written in the Spanish language. The paper was yellow but the printing quite legible, the word *Jesu* appearing many times. None of us could read Spanish, though my mother had studied Latin. We thought the pages were probably from a Bible or prayer book. Unfortunately I took my treasure to school a few days before the term ended for the summer, forgot to bring it home and it was lost forever."

A Bit of Silver Salvage

News stories giving publicity to the Shell Development Company's dating of the beeswax sparked the interest of Mr. and Mrs. Harry Tuttle of Winthrop, Washington. "It will be hard for you to believe," they wrote to Alex Walker, curator of the Tillamook County Pioneer Museum, "but we own a small silver container, taken from the beeswax ship in 1898."

According to the letter, Harry's father, John E. Tuttle, built the first telephone system in Tillamook and his mother was one of the first telephone operators. As a baby Harry quite literally cut his teeth on Nehalem beeswax.

In 1898 when Harry was four years old, his father drove the family to the Nehalem Spit by team and buggy, on an outing made to take advantage of an extremely low tide. Sticking from the sand at low tide line like old driftwood were two weather-beaten masts remaining from some old shipwreck. According to a map drawn later by Harry Tuttle, the location of the wreck was just west of the Nehalem Spit almost at its southern end.

The Tuttles had brought along some shovels so they could explore the wreck. They dug out beeswax both plain and with inscriptions. They also dug out a solid silver container.

The letter revealed that Harry Tuttle and his wife still valued the container so highly they kept it locked in their safety deposit box. Although sixty-three years had passed since its recovery from the wreck they were still curious about its origins.

In 1941 they had sent the container to the Smithsonian Institution, where it was examined carefully. The Smithsonian experts believed it to be a condiment or spice box. They said the hallmark on its lid was that used for silver in several European cities but particularly in Utrecht, Holland, about 1750. They believed the container made about that time, and said that though it resembled Colonial Spanish silver it was more likely the work of a Dutch silversmith.

From a description Mrs. Tuttle had heard of a holy water container she wondered if the silver jar might be such an item instead of a spice box.

Following some correspondence the Tuttles graciously sent the old silver container to Tillamook to be inspected by those in the area most familiar with the history of the Oregon coast. No one in Tillamook was able either to add to or detract from the Smithsonian evaluation, but while the interesting old piece of silver salvage was at hand Burford Wilkerson made some excellent photographs of it, one of which appears in this book.

A Couple of Old Chinese Coins

While walking a lonely stretch of Oregon coast beach in 1955, Dan Lawrence of Tillamook found a curious old Chinese coin. His mother, Mrs. Loran Lawrence, wondering if it might have some connection with an early-day shipwreck, consulted various authorities and numismatists in an attempt to date the coin.

Dr. Thomas O. Mabbott, professor at Hunter College and former editor of *Numismatic Review,* wrote her that the coin bears the name of Ch'ien Lung, emporer from 1736 to 1796. "However, that was a glorious reign and may be no more significant of date than a picture of Washington or Lincoln with us." A curious feature of the coin is a triangular central hole, which some experts believe indicates the piece was struck as a medal, token or charm rather than as money.

Later Mrs. Lawrence was sent a clipping which told that Howard Humphrey, living four miles south of Junction City, Oregon, had found a bronze coin the size of a quarter while digging in his tomato patch. The Oriental Museum determined it to be Chinese, and from two of the characters on its face, of the Ch'ing Dynasty of the Manchu period. Howard Humphrey, Jr., was quoted as saying that "this dates the coin between 1736 and 1796" – the reign of the same Ch'ien Lung.

The Chinese coins found among the Clatsop Indians in early days, when purchased by Judge James Wickersham and submitted to a Chinese scholar for study, were found to bear "the characters of an Emperor who reigned from 1736 to 1799." Again, this would have to be the same Ch'ien Lung, whose reign is usually said to have terminated in 1796. (The *Encyclopedia Brittanica* sets both dates back a year, giving the reign as from 1735 to 1795.)

The Clatsop Indians claimed to have obtained the coins as a result of a shipwreck near their village. Judge Wickersham believed the wreck occurred about 1750. Is it too fantastic to speculate that the two Chinese coins recently discovered, one on the beach and one buried in a garden, might have come from the same shipwreck?

Photographs of both sides of the coin found by Dan Lawrence are reproduced in this book for the benefit of interested numismatists and historians.

Reports of Other Oddities

South of Newport, Oregon: A rusty anchor chain was discovered in 1961 by the crew of a fishboat off the mouth of Beaver Creek. On excavating fifteen feet of the chain, Jim Earl and Bob Cooper came upon the rotting timbers of an unidentified wreck nearby.

Yaquina Bay buoy: An old stock anchor pulled up by a trawl net was believed to be the anchor broken from the chain found south of Newport a few months earlier. Captain Hugh Shaw, a Portland marine surveyor, said, "It's definitely from an old sailing ship. Its ancient stock, shank and fluke design clearly indicate this. The deep penetration of rust indicates that it has been on the bottom for many decades."

Garibaldi: In the 1930s some high school boys went up on the hill facing the town to chop away a snag which had fallen over a big wooden "G" erected to advertise their school. In so doing one of the boys hacked a large brass key out of the tree trunk. Apparently it had been laid in the crotch where the limbs joined the trunk and over the years had become imbedded in the wood. The find was referred to as "the Spanish key" (though it certainly could have been Portuguese or any other nationality). The boy whose property it became is now dead and the key itself lost.

Nedonna: A man living at Nedonna, about a mile south of Manhattan and one-fifth of a mile inland from the present ocean shore, discovered the knee of an old ship. It is pictured in this book as displayed by Ray Lindsay, a Garibaldi photographer with a special interest in windjammer wrecks. The knee is thought to be that of the Cape Horner *Emily Reed*, wrecked at the turn of the century. There is no way of knowing how many bones of unrecorded wrecks lie in these shifting sands where the mouth of the Nehalem River once ran.

Manhattan: Don Viles, editor of the little Garibaldi publication called *The Buccaneer*, tells that a few years ago he was talking to a young girl in a local restaurant who said that her uncle had found some gold coins on a hillside near Manhattan. He had tied a goat to a bush and returned to discover the unruly animal had pulled up the whole bush. In the hole where the roots had been was a container holding the coins. (The location here jives very well with the place the Portugese sailors are supposed to have buried the teakwood boxes; unfortunately Viles does not know the name of the girl or her uncle — otherwise it might be possible to learn if the coins could have been Portuguese.)

Viles also says that an old, full-blooded Nehalem Indian recently told him that about one hundred years ago his stepfather was walking where Manhattan is now and found men digging huge candles out of the edge of a lake and stacking them like cordwood.

The Search of Tony Mareno

In the late 1950s Milo Merrill, heir to Pat Smith's secret lore, gave up his hunt for the Neahkahnie Treasure — only temporarily, as he hoped. He went mining in Nevada and his present whereabouts are unknown.

For a number of years after Merrill's departure from the Oregon coast scene, large-scale treasure hunts were in abeyance. Then late in 1966 the Neahkahnie Treasure once again became big news, this time when a Salem, Oregon, house painter, Tony Mareno, publicly announced he had solved the riddle.

Oddly enough, Tony Mareno's basic orientation turned out to be the same as that of his predecessor, Milo Merrill. Both were highly religious. Merrill revealed to me in a private interview that he hoped for divine guidance in uncovering the fortune in jewels and gold, which he would then put to Christian uses. And Tony Mareno told reporters that he found his clues and inspiration in the Bible. He talked of his search in terms of a religious mission and said he felt he was being led by God.

In the publicity attendant upon his treasure adventure, Mareno was revealed as an interesting man who built fine ship models in his spare time. Good-looking and dark-eyed, he was just past thirty-five years of age when he first began researching the Neahkahnie Treasure. Undaunted by the responsibilities of a wife and growing family (seven children at that time) he had been a treasure buff for years and had made one investigation in Arizona but no dig.

Contradictory statements have appeared regarding Mareno's conclusions about the treasure. According to one account he believed the Neahkahnie Treasure consisted of sacred religious objects sacked from Spanish churches in the 1700s. The loot, estimated at $20 million, was put aboard a galleon which was later hijacked and wrecked on Manzanita Beach at the base of Neahkahnie Mountain during an extremely low tide. The treasure was then placed in coffins and buried on the beach and covered over with a layer of rocks and beeswax.

But in a different account Mareno is quoted as saying: "My theory is that in the late 1500s there was an argument between the Catholics and Spain and they (the Catholics) began pulling all their caches out of their missions, bringing them to America and burying them along the coast." And in this version the value of the treasure jumped astronomically: "Based on my research, I had estimated the treasure at $180 million, but a freelance writer who says he's done extensive research says he believes it's $500 million." From then on the newspapers referred to the "$500 million treasure."

The stone markers found on Neahkahnie and in the vicinity Mareno believed to have been carved by a monk and to bear codes from certain Bible verses. His theory was that the freebooters, in leaving the glyphic stones, were inspired by the first verses of Deuteronomy 10:

> 1 At that time the Lord said unto me,
> Hew thee two tables of stone like unto
> the first, and come up unto me into the
> mount, and make thee an ark of wood.
>
> 2 And I will write on the tables the words
> that were in the first tables which thou
> breakest, and thou shalt put them in the ark.
>
> 3 And I made an ark of shittim wood, and
> hewed two tables of stone like unto the
> first, and went up into the mount, having
> the two tables in my hand.

A large stone, bearing among other signs and symbols, the letters "BKS," which has long been displayed at Manzanita, led

Mareno to the Bible, to Kings I, Chapter 7. This deals with "The Building of Solomon's house, of the house of Lebanon, of the porch of pillars, of the porch of judgment, of the house of Pharoah's daughter, Hiram's work on the two pillars, of the molten sea, of the ten bases, of the ten lavers, and all the vessels," etc.

Mareno stresses the following verses:

> 49 And the candlesticks of pure gold, five
> on the right *side*, and five on the left,
> before the oracle, with the flowers, and
> the lamps, and the tongs *of* gold,
>
> 50 And the bowls, and the snuffers, and the
> basons, and the spoons, and the censers
> of pure gold . . . .

So guided by Scripture he decoded to his own satisfaction the marked stones which have so long puzzled Neahkahnie Treasure fans. He said he located the treasure site in November 1965, but spent the next year doing further research, checking and rechecking, trying to poke holes in his theory. Finding no contradictions, he then used both a metal detector and a small tractor to narrow down the area of search and determine the character of the underlying rock. All set, he made a formal request to the Oregon State Land Board for permission to conduct a serious treasure dig on the publicly owned beach at Manzanita.

Red tape barred the way for months. The State Land Board was not opposed to granting a permit, but a long hangup ensued because the University of Oregon, whose task was to protect potential archeological sites, must concur, and University President Arthur S. Flemming wouldn't say yes and wouldn't say no.

In the end it took a special legislative act to clear the way for the dig. A bill introduced by Representative Jack Anunsen, Republican from Salem, giving jurisdiction in this case to the Land Board only was passed by the legislature and signed by Oregon Governor Tom McCall in April 1967. The "Tony Mareno Treasure-Trove Law" proved a break for the Salem house painter since under its provisions the Land Board would receive only 25% of any treasure unearthed as against 50% claimed by Oregon under traditional state laws.

Immediately on receiving his permit Mareno arrived at Manzanita Beach with a house-trailer, a tractor, a pump, some casings and other equipment. On his heels came photographers, newsmen, TV cameras, freelance writers and crowds of the curious. State Land Board clerk Dale Mallicoat stood by to receive Oregon's share of any treasure found. Mareno became harassed by overattention.

He began his dig, sinking a wooden caisson to keep sand from washing back into the excavation. But though he had believed he would find the treasure within a few days it proved illusive. Early in the summer he was joined in the search by a seventy-two year old retired miner from Seattle, Harrison Bagley.

At fourteen-feet down water boiled into their shaft—"reddish and stinking with some rotten wood in it." With promises of shares in any treasure found, Mareno then enlisted the aid of Ray Daniels and Karl Trussel, Salem contractors, who came with a bulldozer, clamshell, power shovel and steel casings six feet in diameter. When at a depth of twenty-six feet the casings collapsed, new and stronger ones were substituted. At forty feet down the workers hit something solid. As one newspaper headlined it: " 'THUMP' COULD MEAN BURIED TREASURE." Jim Hansen, a diver from St. Helens, Oregon, was called in. Twice he descended into the shaft and chipped pieces of hard material from the bottom.

Mareno pronounced this to be soapstone and said he and his partners would drill through it. But Herbert G. Schlicker, a state geologist, inspected the site and declared that the diggers had hit bedrock—a sandy shale of the Astoria formation laid down fifteen million years ago and undisturbed since then.

At which point Mareno decided his surface X must have been misplaced a few feet and he would have to go laterally through the sand to find the treasure. Always he needed funds and various interested people put up money or services in exchange for his promise of a share in the treasure-trove. In January 1968 the State Corporation Division decided this constituted a sale of securities which must be registered with the corporation commissioner. Mareno complied with commission requirements and requested to sell more shares. An authorization to raise $6000 more was granted with the proviso that buyers of stock be warned that the State Department of Geology believed he had hit bedrock and that his lease on the beach would expire April 19.

When Mareno asked for an extension of the lease, Manzanita property owners made loud protests, complaining of noise, danger, and scenic damage to the beach. A five-week extension was granted but with the requirement of such high performance and liability bonds that Mareno shortly thereafter gave up the search—temporarily, he said.

During his year of excavation Mareno did retrieve a few interesting items from the sand. These included an object he stated was pure silver, a marlin-spike off some old vessel, an old-fashioned wooden spike with writing on it, some links of an iron chain, a chunk of cement, and pieces of a container Mareno called a wine cup. The cup fragments, he claimed, were identified by a University of Oregon scientist as "handmade in either Barcelona or Greece in the 1500s."

Oregon coast dwellers who feared Mareno might return to the beach waxed highly indignant. The treasure search had turned into a mechanized, commercial enterprise, they said. Future digs on state-owned land should be confined to hand tools.

The news that Mareno had given up, packed his things and gone to Canada came from the Washington (D.C.) *Post* in December 1968. The article, written by Harold Hughes, stated that Mareno's alleged silver object had been identified as salt-water preserved iron by the Oregon State Department of Geology. Many Oregonians were glad to see the last of Mareno, Hughes added.

Treasure Signs and Symbols

Since Tony Mareno's interpretation of the glyphic rocks according to Scripture failed some other method of decoding must be considered. In this connection a letter from Ralph Gironda, Jr., of New Rochelle, New York, is of interest. He calls attention to a chapter, "Signs and Symbols," in J. Frank Dobie's popular treasure book, *Coronado's Children.* Since some of the treasure symbols whose meaning are given by Dobie do appear on the marked rocks at Neahkahnie a study of the chapter may be of help in solving this mystery which over the years has grown more difficult rather than less so.

THAT ILLUSORY BLUE BUCKET MINE

FACT AND FICTION OF THE BLUE BUCKET MINE
(Hult, *Lost Mines,* pages 46-62)

This most celebrated of all Pacific Northwest lost mines had its beginnings in late August 1845 when a train of emigrant wagons, attempting to cut 200 miles off the tried-and-true Oregon Trail, ascended the Malheur River and became hopelessly lost in eastern Oregon. Their guide, Steve Meek, mountainman and fur trapper, brother of famed Joe Meek, claimed to know a good route through this way but he became as confused as everyone else. There followed a six weeks' erratic course through mountains and deserts, of such hardship and privation that twenty people died and another twenty were fatally ill by the time the train reached safety at The Dalles.

During this bewildering journey someone picked up a piece or perhaps several pieces of an odd metal. Though a sample was pounded out on a wagon tire and found to be malleable, still no one could guess what it was. Later, on seeing California gold, the members of the wagon train came in for a shock, realizing they had handled some soft nugget gold.

As early as 1849 a search party went back looking for the place of discovery. Twenty-one-year-old Dan Herren, traveling west with the large family of his uncle John Herren, was credited by many of the pioneers with actually having picked up the gold while out looking for lost cattle, and in 1854 Dan himself returned over the trail trying to identify the site, but without success. A number of large parties went out in the early 1860s, most of them being driven back by the Indians, though in 1861 the Adams-Griffen party discovered the rich mines of Canyon Creek in the John Day country, which some believe to be the gold found by the Meek train.

In early acounts of the mine there is no use of the name Blue Bucket, the prospect being called the Lost Diggins, the Lost Emigrant Mines or the Meek Cutoff Mines. Around 1885 a fanciful tale was added crediting the discovery to a little girl playing along a creek with her little blue bucket, which she innocently filled with nuggets. From then on Blue Bucket stories proliferated and many different tales evolved about who discovered the gold and in what area. Over the years search parties have poked into every conceivable corner of eastern Oregon looking for the lost gold.

According to a persistent story, a woman in one of the wagons had died not long after the train left the Oregon Trail. Her grave had been marked with a stone, which if it could be found, would constitute a landmark, because the gold had been picked up one day's travel (or in some versions, two days' travel) west from the site. "Find me the grave of the woman," said Blue Bucket searchers, "and I'll show you buckets of gold."

A gravestone located on the North Fork of the Malheur River just above the Agency Valley Reservoir, scratched with the legend "Mrs. S. Chambers, Sept. 3, 1845," was identified in 1950 as definitely having been left behind by the Meek train. Yet a search west from that site revealed no gold. Could it be that after burying Mrs. Chambers the train turned *south* instead?

A Nugget in a Mole Hole

One of the reasons why the Blue Bucket story has become downright aggravating in its contradictions is that just about every one of the 1000 people who lived through the Meek debacle along the Cutoff in 1845 later told a big tale of *his* part in the gold find—and seldom two of these tales alike. Babes in arms—and even babes unborn—grew up to swear they were on the spot with their little blue buckets, gaining future immortality by picking up the famed nuggets.

Memory is a faulty thing, as most of us know who have advanced into or past middle age. Too often we have found ourselves sure of something which family, friends or written records can demonstrate happened quite differently. Some Blue Bucket stories are no doubt a result of such memory distortion. But the sheer volume of the variant yarns suggests a kind of mass wish-thinking.

As the rather tame story of the Lost Emigrant Gold blossomed into the wildly romantic Blue Bucket legend, members of the 1845 migration sought back in their minds to remember the circumstances of the gold discovery. Few there were who couldn't recall something which clicked: some happening at some vague camp, talk overheard, yellow rocks in abundance, or God knows what. So, seizing on some such item of personal recollection nearly every pioneer in the Meek train built his own towering edifice of a self-proclaimed, self-believed Blue Bucket story. Such tales weren't conscious fabrications, not out and out lies; they represented satisfying myths which placed the teller in a central, important position within the structure of the legend and so enveloped him in Blue Bucket glory.

And what with the natural multiplication of children, and of children's children, there is now a veritable army of people who claim "authentic" knowledge of the Blue Bucket find.

If anything my earlier account of this noted story erred on the side of oversimplication. This was not for lack of reading, pondering and absorbing the myriad claims that run off in all directions into the eastern Oregon landscape; it was that having become totally bogged down in contradiction I was forced to cut a path to daylight with a sharp machete. For when one can read that the gold of the Meek train was picked up by a woman, by a little girl, by teenage girls, by young men, by old men, by middle-aged men; that it was kept, was thrown away, was put in an apron, was put in a blue bucket, was put in a tool chest, was lost in crossing a river, was retained and made into jewelry — well, one begins to think he had better forget the whole thing and take up stamp collecting.

But obviously I could not take up stamp collecting. I was (and am) totally "hooked" on writing; and besides I thought I saw glimmers of hope on the far edge of the jungle. I felt if one could go back far enough toward 1845 to the days before the legend had stirred so many minds to their wondrous inventions, then one might find the simple outlines of a not-too-confusing story. Even if one were able to pan only a few colors, a few grains of truth, these would be of more value than any number of mythical buckets of gold.

So ruthlessly hacking aside all later growth I sought the earliest possible sources. I became convinced that young Dan Herren picked up the modest nuggets which gave rise to the lost mine story and that the Blue Bucket name and claims were of later origin.

A statement which has since been published by a granddaughter of John Herren, Flora Herren Bailey, bears out my conclusions.* Her father Levi Herren was only ten years old the summer his parents came west, but Flora also heard the story from the older Herrens involved, including Dan himself; and what they told her was firmly implanted in her mind because she answered the dozens, even hundreds, of letters of inquiry about the Blue Bucket discovery which were received by the Herren family.

According to Mrs. Bailey the campsite in question was "on the head waters of the Malheur river or one of the small streams that form it", and on a bank above the creek sat several small stone huts. (These are further described in my chapter, "Ancient Tree Blaze, Stone Huts, and Other 'Spanish' Mysteries.")

"It was along this stream," Mrs. Bailey says, "while rounding up the stock to move on, that my father's cousin Dan Herren saw a piece of yellow metal shining in a mound of fresh earth thrown up by a mole or gopher. It was oblong in shape, a little longer at one end than the other and about the size of a man's thumb."

It was three years later before he realized that the piece of metal was a gold nugget. "He went back with a party (Uncle Noah [Herren] was one of them) and would probably have reached the spot but they were driven back by *the Indians.*

"Many romantic stories have gained circulation . . . . the . . . .

---

* This appears in Keith Clark and Lowell Tiller's *Terrible Trail,* to be discussed later in this chapter. Quotations by permission of the authors.

most popular was that the children filled the blue bucket with nuggets and carried them to camp to play with. But Dan Herren says all these were fabrications, and all the golden stories started with *the one nugget picked up* in a mole hill." (The italics are Mrs. Bailey's.)

But just to keep the contradiction quotient high, a different branch of the Herren family has a quite different tale to tell about the gold discovery. We shall have occasion to examine the other version later.

Mr. Meek's Vast Lake

And of course I have long since discovered that the use of early source material does not solve the Blue Bucket problem either. Early-day items can be obfuscating instead of clarifying. I recently obtained an article which appeared in the Eugene City *Oregon State Journal* a hundred years ago. Showing a date of November 16, 1867, and headed "Lost Emigrant Mines," it reads:

"It has been reported that an emigrant train which came the 'plains across' in 1845, discovered, or rather passed through, a country fabulously rich in gold — that nuggets lay around loose upon the ground. This has been doubted by some, and by others full faith and credit has been given the story. However, Stephen H. Meek, brother of the celebrated 'Joe' Meek, knew that such was the case, having been pilot of the train. . . . and for two years has been hunting the locality and has found it.

"Of course the story as told by some was exaggerated, but Mr. Meek found nuggets upon the surface, intermixed with quartz rock, worth from one to eight dollars. He prospected six feet below the surface, and found dirt which paid ten dollars to the pan, although the bed-rock had not been reached. He is satisfied from his discoveries that nothing in California, Idaho or any other country ever equaled these mines. It is the intention of Mr. Meek to organize a company in the spring, of his own choosing, and return to the diggings.

"One great peculiarity of the section of country in which these mines are is the disappearance and re-appearance of a vast lake, sixty miles in length. In 1833 Mr. Meek passed over the identical country, and found the magnificent sheet of water, and actually caught fish and trapped beaver from the same. He was with a train belonging to B.L.E. Bainville [Bonneville] & Co., fur traders of New York.

"In 1845, when piloting the emigrant train through the wilderness, he found the exact spot where this lake *had been* – but where no lake then existed. The train of emigrant wagons passed over the spot, which a few years before had been covered by a magnificent sheet of water, and followed through the valley left by the same, sixty miles, to what had been the opposite end of the lake.

"In 1867, Mr. Meek again returned to the same spot, and imagine his amazement to find the lake as he first witnessed it in 1833 – there before him lay the apparently same magnificent sheet of water, looking as beautiful as it did before its entire disappearance. Now, there can be no doubt of this singular fact; we can produce the evidence of Mr. Meek's own observation.

"Doubting his own senses, he searched the country over until he found the wagon trail made in 1845, and tracked the same to the end of the lake, where it entered the same (the old track, of course, being covered with the water of the lake mentioned.) He then followed the rim or beach of the lake around to the opposite end, (sixty miles), and there found the wagon trail making its appearance from under the water, and, where twenty-two years before the emigrant train had passed over the self-same spot, now covered with water.

"This, to us, and perhaps to the reader, may seem a strange story to relate – but our old citizens know Mr. Meek, and know they can rely upon his statements. Although now fifty-eight years of age, Mr. M. looks hale and hearty, and much younger than his brother 'Joe' Meek, who is his junior."

What I like about this article is that it revitalizes both Meek and the lost mine, taking one, as it does, behind a hundred years of accretions, to the fresh young story.

And what I find maddening is its lack of any kind of clues. For where, pray tell, was the scene of Meek's find, with "nuggets upon the surface, intermixed with quartz rock. . . ."? It could be on the moon.

As for Meek's great lake, possibly he was misquoted and meant to say that in the Harney Valley, along the several lakes there, water had so receded that the wagon train passed over margins which formerly had been inundated and subsequently in wetter times again became so. Certainly it is impossible to believe a lake of the dimensions given could completely disappear and reappear in the space of a few short decades. If this is what he was claiming, then it becomes obvious Steve Meek should have confined his drinking to the lake waters he so eloquently describes.

A Field of Yellow Rocks

The correct name of the man who headed the 1860 search for the lost diggings, I have learned, is Henry Marlin, not Martin; he has been confused with a wagon master in the 1845 train, H. D. Martin, whose first name was Hardin.

Also in the Meek train was Zachariah Moreland, seemingly the Moreland who headed the 1861 expedition. New light has been thrown on this search party by a pioneer, Jacob Manley Currier, writing from Corvallis, Oregon, in 1919. "Marland," as Currier spells it, had been with the 1845 migration and claimed to have seen all kinds of nuggets at one camping place. In July 1861 he led out a group of forty-four men from Lane County, among them Eugene Skinner, founder of Eugene, Oregon; Kenney, an old prospector (possibly the same J. Keeney, who reported gold prospects in northeast Oregon that same fall); Kenney's (or Keeney's) son; Oren Belknap, Bill Montgomery and Currier. Members of the party did a lot of prospecting all the way from the Deschutes River to the mouth of the Malheur.

But Currier became disillusioned with the search and explains why:

"We found no gold but the section where Marland saw the nuggets was full of yellow stones that resembled gold. One day Marland and myself were riding along together. He asked me if I did not think that a man 16 years ago might have taken those yellow stones for gold. So far as I was concerned the expedition failed right there. . . ."

We need have no fear that the whole Blue Bucket hullaballoo rests on those yellow stones, but this does seem a good example of the towering edifice built on an authentic but wishfully misinterpreted recollection. After hearing the glad truth about the find along the trail, Moreland began to rack his brain and in remembrance certain rocks began to glow. Eureka, the gold! And after toiling back to the scene at the head of a party he was an honest man, even an imprudent one, to admit even haltingly that his memory might have played him false. Men have been hung for less.

But the majority of the 1845 pioneers never had occasion to test their favorite theories in the field and so never were assailed by doubts.

On Stealing the Blue Bucket

I have been quite amazed to discover one variant yarn which claims that the famed lost mine does not lie in Oregon at all – its true location is in Nevada.

In this version the big Oregon migration of 1845 did not travel by way of the Malheur, for it had left the old Oregon Trail long before it reached Fort Boise. The wagons kept together along the California Trail until on the Humboldt River near the present town of Beowawa, Nevada, the Oregon contingent split off, turning northwest into the Black Rock Desert, crossing the Black Rock Mountains, and with High Rock Canyon as significant landmark, found gold enough to fill the wondrous Blue Bucket.

The route described is that of the Applegate Trail, the southern route into Oregon, but unfortunately for the Nevada claim this trail was not opened or used until the following years of 1846.

Besides that there is a mountain of hard evidence proving that Fort Boise on the Oregon Trail was the place Steve Meek persuaded the men in 200 wagons to pay him $5 a wagon to lead them along the proposed cutoff, and that it was from the Snake River not far from Fort Boise that Meek led the party west from the Oregon Trail. Not one of the pioneer stories ever placed the Meek fiasco anywhere but in eastern Oregon; all the contradictions arise as to where within those limits the train went in its wanderings.

How then could the Nevada claim possibly be made? Harold's Club of Reno once put forth the story in a promotional booklet

Al Reiner, for ten years a packer and guide in the upper Willamette country of the Oregon Cascades, took many treasure hunters up to Bristow Meadows, high in the Calapooya Mountains, and in due time developed his own theory about the Two Frenchmen Mine.

Left: Tree blaze, discovered inside the butt section of a dead tamarack tree cut about fifteen miles north of Bonners Ferry, Idaho, in 1936, by Leon A. Patten and estimated by several observers to be more than 200 years old. *Right:* Cover piece of butt section of the dead tamarack tree which, when split off, revealed the tree blaze.

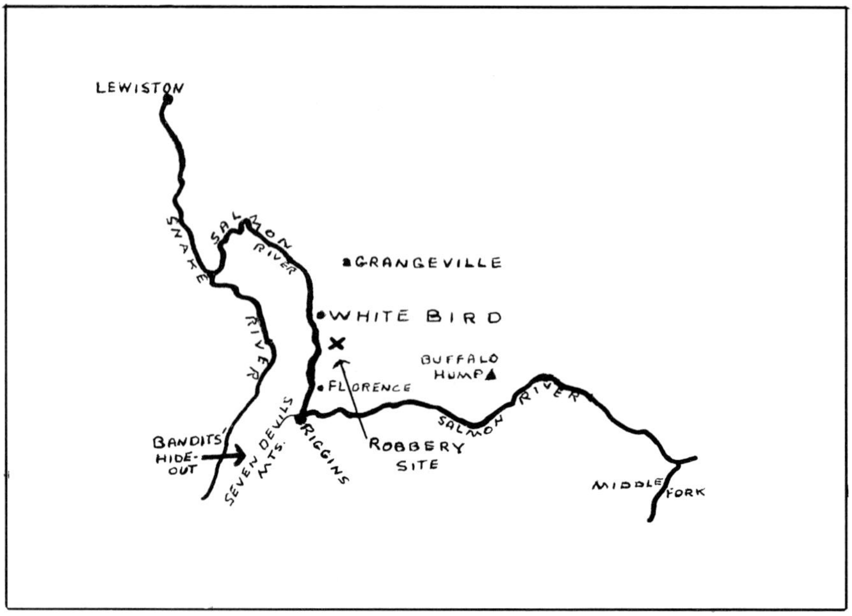

SALMON RIVER CACHES

According to legend, two caches are buried somewhere near White Bird, a town in the Salmon River country of Idaho. One consists of "lots of gold" hidden by an Indian named Black Tail. The other is a hoard of $75,000 in gold dust stolen from miners en route from Florence to Lewiston. The robbers, being pursued, secreted the loot among the rocks along the Salmon River.

The scene of this robbery was on the east side of the old pack trail along the Salmon River south of White Bird. After concealing the gold, the bandits headed for the rough Seven Devils area, but all of them were eventually overtaken and slain before returning to find their cache. There can be little doubt that this fortune still remains among the rocks in the canyon. (See page 198)

Among mines that have been most persistently hunted is the Lost Cleveland on the Middle Fork of Salmon River. A man named Cleveland followed a rich float up the mountainside a half century ago and uncovered a ledge of very rich ore. He gathered all that he could carry and sold it to the mint in San Francisco and then returned to his home in Missouri to reflect on his good fortune and to visit his relatives. The next year and thereafter he was unable to find his ledge of gold. He died while still searching and today his body lies near the Yellowjacket Mine in Lemhi County. The Lost Cleveland Mine has lured many prospectors into the wilderness of the Primitive Area but it has never been found.

and at first I was inclined to blame it on some facile public relations man. But I find that the Nevada story is much older than that particular booklet.

After the Applegate Trail was opened in 1846 it was the route used by many wagon trains making their way to Oregon. The story may rise out of something which happened to a later emigrant party traveling through the Black Rock Mountains and on toward Oregon – some event which as time passed became confused with elements of the Blue Bucket legend, either through ignorance of the true history of the Blue Bucket or through a desire, conscious or otherwise, to appropriate some of its glory.

At any rate nothing so proves what a great treasure tale the Blue Bucket legend is as having a neighboring state try to steal it.

This Story Was Only a Gasser

You can hardly argue with a man who was on the scene, and R. W. Pitney of Junction City, Oregon, has written me that one of the Blue Bucket searches I mentioned simply never happened.

The mention he refers to will be found in the story of the "Four Dutchman Mine" (Hult, *Lost Mines,* pages 111-112), which tells of some Dutchmen who wandered up from California in gold rush times and found a rich placer mine in a canyon thought to lie in the Ochoco Mountains. They left at the end of the summer, never to return.

In the Portland *Oregonian* there appeared an item about two Baker, Oregon, men, W. J. Buckley and K. C. Harpin, who in 1916 went out hunting the Blue Bucket Mine and in the Prineville area came upon an old sluice box and cabin they believed to have been left behind by the Blue Bucket search party of 1849. But since most of the early searchers headed for the Malheur country and did not linger long enough to build a cabin, I speculated that this might have been the camp of the four Dutchmen.

In his letter Pitney puts these down as "strange statements" and goes on to say:

"Joe Buckley owned the Mint Saloon in Baker and I know he was not a member of the party you mentioned. I was a friend of Tom Gorman, and in 1916 I met him and Kit Harpin near Robbers Roost, some two miles from the ghost town of Auburn, Oregon. They were returning from a short trip but they were certainly not looking for the Blue Bucket Mine. The only thing they found was an old wagon tire with a small tree growing up inside it. That night in the Mint Saloon, Mr. Gorman, being a wonderful storyteller, facts or otherwise, spun a great tale about the two finding the Blue Bucket placer mine. He later told me he had noticed a cub reporter from the *Morning Democrat* and told the tale simply to pull the reporter's leg.

"But the story was written up and published in many coast papers. Mr. Gorman even received a telegram from a man in San Francisco whose father had been a member of the wagon train of Blue Bucket fame. He wanted to finance the venture but had to be told it was all a practical joke. The newspaper owner was a man who had grown up in the mining country and should have known better than to publish the story as news."

Touche! That explodes one Blue Bucket story.

Discovery of the Boise Basin

It is sometimes said that the Boise Basin was discovered by a party looking for the Blue Bucket diggings, a statement which would indicate that the lost mine search was carried as far east as Boise County in southern Idaho. Here is what really happened.

Moses Splawn was the moving spirit in the Boise Basin adventure. He had been mining in Elk City, Idaho, and in the summer of 1861 rushed down to Florence when gold was discovered on the Salmon River. Several times he encountered a Bannock Indian who evinced special interest in the mining activity and told Splawn that "in a basin in the mountains far to the south, he, as a boy, had picked up chunks of the yellow metal" — the same stuff that was causing such a furor among white men. He described the basin so vividly that Splawn felt he could recognize it.

In the spring of 1862 Splawn decided to go on a search for the golden basin but he considered the Indians so dangerous that he must round up a large party of men. With a few friends he went to Auburn, Oregon, a boom town on the site of the Canyon Creek discovery. There he found Captain Tom Turner and a party of fifty men who had come from the Willamette Valley with plans to search Catherine Creek, a little north of Baker, for the Blue Bucket Mine. It was agreed that Splawn and his men would first help the large party look for the Blue Bucket and if that search proved unsuccessful the combined party would then follow Splawn's directions in trying to locate the basin.

The men did find a gold digging on Jordan Creek (near where Silver City and Ruby were later to be located) but were forced to admit defeat as far as the Blue Bucket Mine was concerned. Splawn then reminded Captain Turner of his promise to help search for the basin, but only seven men would follow Splawn. Near Fort Boise this small party met another small party led by George Grimes and they joined forces.

They soon ran into hostile Indians and some of the men dropped out. Those who persisted were Splawn, Grimes, Jack Westenfelter, Joe Branstetter, D. G. Fogus, Jack Reynolds, Wilson, an unnamed Englishman and four Portuguese. They located the basin and on the creek where Centerville, Idaho, was soon to be founded, discovered rich placer gold. George Grimes was almost immediately killed by the Indians and the other white men forced to retreat until they could return with a stronger party.

Splawn knew he had triumphed: "I felt I had found the basin of my dreams. The story of my Indian friend was true."

But though this first major gold discovery in southern Idaho was only indirectly related to a Blue Bucket search, such was not the case in connection with the second important discovery, made 120 miles to the south.

In the spring of 1863 a party of twenty-seven miners set out from Centerville, a Boise Basin boom town. These men were definitely in search of the "lost diggings" but for some reason did not head toward the Malheur. Instead they kept to the south side of the Snake River and upon climbing a divide saw southwest indications of a large river which proved to be the Owyhee. They struggled on to its headwaters, believed at the time to lie in Oregon, and along the same creek where the Turner party had dug they found colors in abundance, They named the stream Jordan Creek for Michael Jordan, a member of the party, and after staking as much ground as they could they hurried back to Centerville. News of their discovery caused what Bancroft calls "a kind of instant insanity," as 2500 men rushed off to the Owyhee. Gold at the site proved disappointing but the tributaries to Jordan Creek were soon revealed to bear wonderfully rich silver ledges.

When Idaho Territory was created its foundations rested on the gold of the Boise Basin and the silver of the Owyhee country. So though no one ever claimed Idaho as a possible location of the Blue Bucket its history was nonetheless shaped by the lure of that lost mine.

"We Have Traveled the Exact Route"

In connection with treasure hunts, blisters and calluses have become things of the past. Lost mine adventurers use cars, trucks, jeeps and even airplanes to scour the hills. Gasoline burned rather than shoe leather worn is the order of the day.

Searching by means of motorized vehicles takes away one kind of experience and substitutes another. Certainly a well-organized, well-equipped expedition on wheels has a real advantage in the wider territory it can cover while checking out clues far distant from one another. And when you consider the long, long trail the lost wagon train traveled you realize that mobility is a necessary requirement for tracing those old tracks.

In April 1958 L. L. Mills and his son-in-law, C. L. Carlson, both of Kelso, Washington, outfitted a station wagon and drove into eastern Oregon in search of the Blue Bucket Mine. In part Mills was inspired by my book *Lost Mines* and in part by Blue Bucket yarns he himself had heard over the years. Before his departure he held several conferences with Ora M. Yeager of Castle Rock, Washington, who in his home town of Heppner, Oregon, had once lived neighbors to two sons of William J. Herren (the oldest son of John Herren) and had often heard their stories of the gold discovery. Yeager had also obtained a written statement from Earl E. Wallace of Medford, Oregon, grandson of Susan Herren Wallace who was with the Meek party in 1845.

Wallace's statement reads in part: "After leaving the mountains they traveled in a northeasterly direction [northwesterly must be meant] toward Wagontire mountain. . . . they were having trouble finding water for themselves and stock, and grass was very scarce. . . . Their only landmark was a Chalk Mountain.

"After two days' travel from the mountain they came to a small creek with potholes of water but not enough. . . . Will J. Herren went to the creek for some water and in one of the potholes discovered some yellow pebbles" – the Blue Bucket nuggets.

It will be noted that absolutely none of the details in this statement corresponds in any way with the statement of Flora Herren Bailey.

Ora Yeager, using the clues in Earl E. Wallace's statement as well as another account mentioning three pine trees near the site of the gold discovery, helped map out an itinerary for Mills.

Mills and his son-in-law began by making a thorough investigation of portions of Malheur and Harney counties. As they went along Mills charted the route he believed the Meek train followed. According to his ideas the wagons crossed the Snake River near present day Nyssa, Oregon, and kept south of the Malheur River for several days. They traveled in a southwest direction past Chalk Butte (not the Chalk Mountain of Earl Wallace's statement), and on to Coyote Wells at the head of the Dry Creek which flows into the Owyhee Reservoir. They continued on south to Skull Springs and passed through the northern reaches of the Steens Mountains.

Here they turned rather abruptly northwest, crossing the southern end of the South Fork of the Malheur and entering Crane Valley. Passing the site of present day Crane they continued on west along Malheur, Harney and Silver Lakes. Seemingly Mills believed what the others have called Stinking Hollows, where the train paused for five days, was at Silver Lake (now dry) – and in spite of Earl Wallace's statement he does not mention their going on to Wagontire Mountain.

Instead he indicates the wagons turned northwest up the

Silver River (or Creek, as it is often called). Crossing into the Crooked River drainage and continuing still northwest the train passed just south of Logan Butte, which Mills describes as a chalk butte visible for miles and used by emigrant parties as a landmark.

And in what must be the Maury Mountain section of the Ochoco National Forest, Mills came upon an operating gold mine which he decided was the real site of the Blue Bucket Mine. It was owned by a Bend man and was paying $7.50 a ton.

"This location fits all the clues," Mills wrote to Yeager, "[being] approximately two days' travel past a chalk mountain. Then one day's travel to the timber, up a steep grade to the summit. There are three pine trees but one has been cut. It is also about two days' travel by oxteam in a westerly direction from an unidentified grave which is near an unnamed spring."

In summing up his search Mills wrote, "We traveled nearly 2000 miles, covering all the creeks and rivers we could find running in a southwest direction, and we are positive we have at last traveled over the exact route taken by the lost wagon train and [discovered] the Blue Bucket Mine."

But Ora Yeager was anything but satisfied with Mills' conclusions. And as we shall see, other parties following other itineraries based on Blue Bucket material they consider just as authentic have sought and even "found" the mine in other locations.

Scientific Blue Bucket Safari

This particular search began with the interest of William Lambert of Portland, Oregon, an electronics expert. In the summer of 1957 he and P. A. Buckmaster, a California miner, spent three weeks tracing old wagon tracks through northern Malheur County, where they are still visible, and finding, as others have done, the grave of Mrs. S. Chambers. The gravestone, now set in concrete, is the only physical monument left behind by the Meek train and is considered by many to be an important landmark in the search for the gold.

These adventures whetted Lambert's desire to know more exactly the route taken by the Meek train, and during the following winter he and a geology student, Robert Scranton, poured over diaries, letters, newspaper items and other records of the 1845 disaster. They were directed in their research by the prominent Oregon historian, Lancaster (Jack) Pollard.

Using his knowledge of geology, Scranton undertook to determine in what areas the Meek train might have passed through goldbearing terrain.

It is well established (in spite of Mills' view to the contrary) that the train ascended the Malheur River from Vale to a little east of Harper, crossed over to Bully Creek, then took a course mainly west to the North Fork, the site of the Chambers' grave. But where did they go from there? Some Blue Bucket buffs, myself included, have believed that the train probably turned generally south, entering the Harney Basin by way of Crane Valley.

But the researches of Lambert and Scranton led them to believe the Meek train kept a somewhat zigzag course almost due west, crossing creeks tributary to the Middle Fork (a stream that on many maps is shown as the upper main Malheur River) before turning south toward the Harney Basin. This accords with old stories that the gold was picked up one or two days' journey *west* of the grave of the woman.

So it was the area west of the Chambers grave on the North Fork that sparked the curiosity of Lambert and Scranton. With spring fever upon them they formed the "Blue Bucket 1958 Expedition," which Jack Pollard joined as historian and Buckmaster as chief prospector. Completing the party were Carthel Lanier and M. E. Hunt, eastern Oregon wilderness camping and hunting experts, Walter Eastman, a mining man, and Steve Roth, a photographer.

The expedition was the most luxurious and well equipped ever to strike out for Blue Bucket regions up to that time. A truck carried an icebox, gasoline stove and lantern, folding chairs and mess table, sleeping tents and air mattresses, as well as a fine array of food and refreshments. Behind the truck was a trailer loaded with a jeep of four-wheel drive. In the way of scientific instruments were an M-scope, mine detector, microscope and Geiger counter, but old-fashioned mining equipment was also included. Good maps, both old and new, were in abundance. Some expedition members traveled by car.

Starting June 1, the safari first stopped in the John Day country at Canyon City. Charlie Brown, curator of the Grant County Museum, when asked his assessment of the expedition, answered, "You're all crazy."

Undaunted, the party continued on down the main highway to Prairie City, took dirt roads in the direction of Drewsey, turned and followed a rough two-rutted trail for ten miles, and set up camp on Otis Creek.

Jack Pollard, having a severe heart condition, remained behind as camptender and record keeper while the other seven men formed into two prospecting parties. Dividing the scientific equipment between them and carrying gold pans, picks and shovels, they went forth to investigate the surrounding territory.

They prospected such tributary creeks to the Middle Fork as Otis, Griffin, Birch, Cottonwood and its many subsidiary streams, Calamity, and even one named Blue Bucket. They explored the North Fork as far as Lost Creek, and in all prospected all creeks named and unnamed in an area of 325 square miles.

To their dismay the prospecting parties discovered that the M-scope and mine detector were useless because of their non-discriminating reactions. For example, the M-scope picked up a rich deposit which turned out to be a rusted iron drum. It was the tried-and-true equipment such as pick and shovel, mortar and pestle and gold pan which proved their value. But after pulverizing, grinding, panning they found no grains, no nuggets, no dust — just a few colors in the form of mere specks and not enough of these to prove encouraging.

"After a week of such hunting," wrote Pollard in a Portland *Oregonian* article, "all agreed that the lost Blue Bucket Mine was not on any of the creeks within the area covered. Historical research and scientific prospecting had failed. The lost Blue Bucket Mine is somewhere else." The main contribution of the expedition, he concluded, was that "it proved where the Blue Bucket Mine is not." And the safari moved on northeast in an effort to find the old ghost town of Cake.

Foley Creek Canyon Theory

Lois Allen Pierce, who lives at Hoodsport, Washington, has deep family connections with the Blue Bucket story.

Her paternal grandparents, James Miller Allen and Hannah Jane Riggs Allen, were a young couple with a two-year-old son, Cyrus, when they set out west in 1845. In the same party were Hannah's mother and father, James Berry Riggs and Nancy Anderson Riggs, also Hannah's twin brother Milton, as well as six other assorted brothers and sisters. On leaving St. Joseph, Missouri, they traveled with the W. G. T'Vault train, of which John Waymire was lieutenant and James Allen one of the sergeants. This wagon train later split into several parts but Mrs. Pierce indicates that the Allens and Riggses were still with the T'Vault train when they followed Steve Meek west from the Oregon Trail.

As a girl Lois Allen heard family stories from her Uncle Cyrus Allen, as well as from members of the Riggs family. She later became a friend to Joseph Packwood, whose parents, Elisha and Paulina Prothero Packwood, traveled the Meek Cutoff, and she also heard stories from another friend, Ed Fox, whose grandfather (name not given) was in the same party.

From all these sources Mrs. Pierce put together her own version of the adventures and route of the Military Immigrant Train, as she calls that part of the Meek train with which her forebears traveled.

She believes the Military train, the Herren train and the McNary train were traveling together about two days ahead of the party which buried Mrs. Chambers and were already well on their way to Crane Valley by September 3.

The McNarys and Herrens were so disgusted with Meek that they pulled out of Crane Valley without a guide and started west toward the Harney Basin. But the Military Immigrant Train, numbering about forty-five wagons, stayed with Meek, who led them south of Malheur Lake, on southwest across the Donner and Blitzen River into what is today called Frenchglen. After a rough time through a bad pass the wagon train arrived at the series of small lakes known as the Warner Lakes, which lie in Lake County where it juts east under the corner of Harney County.

The wagon train went into camp near Blue Joint Lake at a place Mrs. Pierce calls "the Marshes," which from her description seems identical with what Samuel Parker called "Stinking Hollows" and others "Lost Camp." Only alkali water was found, which caused anxiety, hardship and quarreling. Among the scouts who went out looking for a better water supply was Rufus Riggs, Hannah Allen's eighteen-year-old brother.

Leaving "the Marshes" the train traveled a little farther south before turning west between the Rabbit and Coyote Hills. After three days they arrived at a canyon trending in a southerly direction, where water was to be had in potholes among the black rocks of the stream bed. They also dug a large water hole in the sand - and while camped in the canyon discovered some shining rocks later identified as gold.

Fall was closing in so the wagon train did not tarry but made a somewhat zigzag course northwest through the desert and on to the Crooked River near the Prineville of today. Still continuing northwest they crossed the Deschutes River and arrived at The Dalles on November 8, 1845. This date is a month later than most of the contingents are *known* to have arrived on the Columbia River.

At the time Mrs. Pierce was charting this route on paper she had not yet visited the eastern Oregon sites under consideration but she was in communication with several people in Lake County, including Mr. and Mrs. Marvin Weatherly of Silverton, and Mr. and Mrs. Lynn Tomlinson of Lakeview.

Marvin Weatherly was first to indicate a belief in Mrs. Pierce's theory. After visiting her home in Hoodsport and listening to her evidence and ideas he offered his help in locating the trail, since he knew the country well. Mrs. Pierce found him and also the Tomlinsons "wonderful."

In his search for the canyon where the Military Immigrant Train had picked up the gold, Weatherly was alerted to look for the only stand of juniper trees for miles around, a stream running south showing black rocks in its bed, and a large water hole where the pioneers had dug for non-alkaline water.

A short time later Weatherly phoned Mrs. Pierce and advised

her to come down immediately – he had found everything just the way she had said he would.

So Lois Pierce, together with her husband D. H. Pierce and their son, drove in hurried excitement to Lakeview. From there Weatherly guided them to Plush and then on northwest into Foley Creek Canyon, where he pointed out the correct juxtaposition of juniper, south-running stream showing black rocks, and a nearby water hole. Mrs. Pierce immediately accepted this as the place where the Blue Bucket gold had been found.

She states that Ed Fox's grandfather said the gold came from a crater in southern Oregon and according to geologists the Foley Creek Canyon is actually a crater. And she is further convinced because several rich pockets of gold have been mined only a few miles from Foley Creek. Still further, if this were the right spot she believed three graves would be found to the north and west, each an oxteam day's journey apart, and sure enough the Pierces located three such graves in the correct locations.

Following their investigations the Pierces, Tomlinsons and Weatherlys all staked mining prospects in the Foley Creek Canyon, calling them the Blue Bucket Group– which means that the lost mine was "discovered" nearly sixty miles south of the route the lost wagon train is believed by most historians to have traveled.

In 1962 Mrs. Pierce published the story of her quest and conclusions in a small book, *Lost Immigrants of 1845 and the Blue Bucket of Gold.* If any of her sources were written ones she does not say so and she cites no documents to give weight to her claims.

She was apparently unacquainted with the diary account of James F. Field, a member of the Riggs train, which has since come to light. As we shall see later, the Field diary tends to demolish Mrs. Pierce's thesis.

Nonetheless Mrs. Pierce was able to interest the staff of *Desert,* a magazine published in Palm Desert, California, in her story, with the result that a well-equipped party made a four-day search through the Oregon desert. Members of the party were Erle Stanley Gardner, writer of mystery stories and treasure tales; Jack and Choral Pepper, respectively publisher and editor of *Desert;* J. W. Black and Dick DeShazer. Reub Long of Fort Rock, Oregon, co-author of *The Oregon Desert,* served as guide. Making their headquarters in the Lost Forest the party sought southeast for the three graves which are prime clues in proving Mrs. Pierce's contentions. These they discovered without difficulty. They then proceeded to Foley Creek, and though they make no mention that Mrs. Pierce had been there before them and had even staked "Blue Bucket" claims, they decided Foley Creek Canyon was the most likely source of the lost wagon train gold—all except Choral Pepper, that is, who held out for Rabbit Creek, the next stream north.

And what with the Blue Bucket Mine being so often discovered and rediscovered, it is no surprise that the *Desert* article reporting the search claimed that the party had "solved" the case of the Blue Bucket gold.

Terrible Trail: A Study of the Route

The significant event of the decade in connection with the Blue Bucket legend has been the publication of the book, *Terrible Trail: The Meek Cutoff, 1845,* by Keith Clark and Lowell Tiller (Caldwell, Idaho: Caxton Printers, Ltd., 1966).

The book represents an exhaustive study of the route taken by the Meek train and if we accept its conclusions it allows us to assess old and new searches, clues, theories, and all those "discoveries" of the Blue Bucket Mine.

Keith Clark is an English professor at Central Oregon College at Bend, and Lowell Tiller a Portland business man. As native Oregonians they had grown up with Blue Bucket and Cutoff yarns and in the winter of 1960 hit upon the idea of a careful examination of the tragic history of the lost wagon train. Their objectives were four:

"To write an investigative if not definitive work on the subject.

"To lay to rest the mass of misinformation about the route of those lost and wandering pioneers who made up the Meek Train.

"To narrate the passage of the emigrant train whose tragic experience rivaled—indeed, surpassed, in the number of deaths —that of the Donner Party of California.

"To help lay the groundwork for the solution of the Blue Bucket mystery."

The authors are definitely not treasure hunters but scholars interested in adding to the general knowledge. To their embarrassment they discovered that questions asked about the history of the Meek party were taken to be questions about the lost mine, and when they denied they were seeking the Blue Bucket people looked at them "with instant incredulous amusement." They admit, "We rather hated to ask, after awhile."

Yet as their formulated objectives show they were well aware of the value to all treasure hunters of a close study of the Meek route. For if searchers knew exactly where the wagon train went they could comb the track foot by foot and inch by inch—and there are plenty of Blue Bucket enthusiasts who would do just that.

Clark and Tiller have done a magnificent job of research, both into printed sources and out in the field. Besides the firsthand accounts I relied on in my 1957 study they discovered others of importance, in particular the diary of James Field, a member of the Riggs train, and the fragment of a diary kept by John Herren. They also used the diaries of Andrew and James McClure, members of the Elliott party which traveled the Cutoff in 1853 and "followed the dim tracks of the Meek party" at the beginning of the route. The text, the footnotes and the bibliography all reveal the authors' extensive sources and make the book a *must* for Blue Bucket buffs.

My one disappointment is that the maps included in *Terrible Trail* show only an approximation of the route the text lays down, and even at times a misleading approximation; they fail to name the important forks of the rivers, such as the Malheur, and also fail to name creeks or to identify the different ranches—even though the locations of these features must be known if one is to follow with any exactitude the route they chart. The reader needs auxiliary maps to understand their theory, which is perhaps not too great a drawback since treasure hunters are—or should be—long on maps.

And where, after all their hard work and study, do Keith Clark and Lowell Tiller believe the wagon train went?

In general the route they believe in does not differ greatly from the commonly accepted one:

—up the Malheur River and across some of its northern branches

—down into the Harney Basin and west along its lakes

—north and west through the Crooked River and Deschutes River drainages to safety at The Dalles.

In particular, however, they pin the route down more closely than anyone else has dared to do—wherein lies their major contribution.

In connection with the early part of the route which has figured so prominently in searches for the Blue Bucket gold, let us note the Clark and Tiller answer to that tantalizing question of which way the Meek wagon train went after it left the site of Mrs. Chambers' grave. The grave, as we know, lies on the North Fork of the Malheur one mile north of the present Agency Valley Reservoir.

From that point the authors of *Terrible Trail* lay down the following route:

—on north up the North Fork toward Castle Rock (called Fremont's Peak by the emigrants)

—southwest to Cottonwood Creek, an easternmost branch of the Middle Fork of the Malheur

—still southwest into the Drewsey Valley, keeping a little north of the present town of Drewsey

—over the hill to upper Pine Creek, a western tributary to the Middle Fork

—south by way of East Cow Creek or possibly Rock Creek, into the Harney Basin.

My own inquiries into the question of the Meek route had revealed even the best sources to be impossibly vague and frustrating. The Parker and Harritt diaries all too often fail to give the direction of travel and the daily mileages they set down conflict sharply. Since the James Field diary sounded so much more definite, I was eager to get my hands on it.

When I wrote to the authors of *Terrible Trail,* Keith Clark and his wife Donna responded by sending a typed copy of the Field diary as well as a duplicated copy of the Herren fragment. In studying this new material I at once noted that James Field in three different entries stated the wagon train was on the South Fork or its tributaries, and John Herren wrote of reaching "a south branch of Malheur." Jesse Harritt had also mentioned the South Fork, but like other historians I had dismissed his lone statement as an obvious error, since the train in coming down from the vicinity of Castle Rock could not have reached the South Fork as quickly as indicated.

With *five* different references to the South Fork having turned up it looked as if a reconsideration of this point was necessary. True, the Meek party was wrong about the names of most of the rivers they encountered. Yet if one assumes they were not on the South Fork then one must ask what branch they could have mistaken for it.

A study of the Malheur River system on the map reveals that it would make perfect sense to call all that part of the river flowing up from the south as far as present day Juntura the South Fork. For it is not until the North Fork joins it near Juntura that the river swings off east. Today we call the southern portion of the stream the South Fork only as far as Riverside but this is an arbitrary decision. Our geography is full of confusion as to what are main rivers, what are branches and how they should be designated.

The emigrant train traveling down the North Fork to its mouth would see it joining a stream flowing up from the south. Quite naturally they would call this converging stream the South Fork. If the train then kept somewhat west of this southern fork, crossing its tributaries as the wagons wheeled south and entered the Harney Basin by way of the Crane Gap—then all the diary entries would be satisfactorily explained.

But when I wrote this suggestion to Clark and Tiller they replied, "Honestly, Ruby, your route may survive in Seattle or Phoenix, but the desert rats of the great basin might want to send you some uncooked camas lily if you persist in it."

They argue that the route south from Agency Valley to Juntura via the North Fork is not geographically feasible since much of it consists of nearly impassable rocky canyons, and that from Juntura to Crane Valley the wagons would have been forced to follow a washboard pattern—surely too rough a course for pioneers who were usually smart enough to stay on ridges. They further point out that this route fails to conform to the descriptions found in the diary accounts, especially in the matter of pine trees. Field on September 4 and Harritt on September 5 (and here they were traveling a day apart) mention pine timber, yet there are no pines and never have been pines where I would place them that day.

In this exchange I was distinctly outclassed, since Clark and Tiller have spent years hiking the trails, jeeping the rough spots and even flying over the canyons and deserts of the eastern Oregon country under discussion, whereas I know it mainly from flat maps. But being determined that all those references to the South Fork should not be ignored, I persisted in spite of the threat of the death camas. But it was not until I was able to meet Lowell Tiller in Portland and debate the matter face to face that we came to grips with our real points of difference. When he showed me some amazingly contoured Army relief maps which revealed the shape of the Malheur country I was impelled to admit that he and Clark were right: The wagon train could hardly have gone south by way of Juntura as I had been contending. The Clark and Tiller route seems far more sound.

Yet this leaves one unresolved contradiction. For if the train passed from the Agency Valley directly to the Middle Fork of the Malheur, then it is the Middle Fork which the 1845 diarists refer to as the South Fork. Yet this stream flows almost directly north to south where they encountered it, a circumstance which makes it strange indeed that they should have called it the South Fork.

Even so, the wagon train apparently did travel west rather than south immediately after leaving the site of Mrs. Chambers' grave and so passed through the area so meticulously prospected by the Lambert-Scranton expedition in 1958 without the faintest showings of gold. Since the Malheur drainage is fairly well eliminated as a possible location of the Blue Bucket Mine, the gold must have been picked up later in the journey. The conclusions of Clark and Tiller as to where along the trail the celebrated event is most likely to have occurred will be of interest to all future Blue Bucket seekers.

Field vs. Foley Creek

Since James and Hannah Allen remained throughout the journey with the wagons of Hannah's parents, the James Rigges, and since James Field was also traveling with Captain Riggs, the Field diary ought to lay down the route Mrs. Pierce contends her forebears traveled.

Yet nothing in the James Field diary agrees in any way with the statements of Mrs. Pierce. Field mentions no such split at Crane Valley as she describes, and a number of things he tells indicate the split did not occur. On September 9, three or four days after the alleged breakup of the Riggs, Herren and McNary trains, Field writes of the burial of a child. A comparison of diary accounts shows that he is describing a burial also attended by Parker and Harritt. Harritt being with the McNary train and Field with the Riggs train it follows that these two parties had not separated.

Field also mentions the Tetherow and Owensby parties, which puts nearly all the segments of the Meek train together, and that Meek himself was still present is shown by the diary entries of September 11 and 14. Parker, in describing the camp on September 16, says that 198 wagons were assembled in one place which certainly must represent the entire company, since 200 wagons are usually said to have followed Meek in the first place.

Field's description of the dry camp where the company remained for some days and from which parties went out looking for water is in essence the same as Parker's and Harritt's. Mrs. Pierce tells much the same story, only locating the camp nearly sixty miles to the south. James Field does not give enough mileage traveled in a southerly direction for the train to have reached the Warner Lakes, and immediately upon leaving the dry camp he notes a north and northeast direction of travel. From the dry camp Mrs. Pierce moves the Military Immigrant Train west between the Rabbit and Coyote hills to Foley Creek Canyon.

James Field became ill and ceased to make diary entries after September 24 but up to that point he is obviously keeping pace day by day with Parker and Harritt, so it is natural to assume he reached The Dalles at approximately the same time they did, early in October, and not the first part of November as Mrs. Pierce would have it.

Since Field continued to travel with the Riggs train (a reference to Captain Riggs on September 16 proves this) and since the pace and route are the same as those of the main contingent, it becomes obvious that the long dip south to Foley Creek Canyon by the forty-five wagons Mrs. Pierce calls the Military Immigrant Train simply couldn't have occurred.

"New Information"

And the Blue Bucket search goes on. Last heard from that indefatigable team of trail tracers, Keith Clark and Lowell Tiller, were off to an eastern Oregon rendezvous with a trailer caravan coming from California with what was claimed to be "new information" about the Blue Bucket Mine. But that obviously, is another chapter. . . .

UP TO BRISTOW MEADOWS

THOSE LOST CABIN MINES (Hult, *Lost Mines,* pages 63-76) .

In southwestern Oregon several different lost cabins are purportly the key to rich mines or hidden loot. A party out from Jacksonville, Oregon, seeking one such Lost Cabin Mine in 1853 discovered an incomparable treasure of a different kind: beautiful, magical Crater Lake.

Another lost cabin is that of two Frenchmen who used to fade into the wilderness off the old Fort Klamath Military Road, somewhere between Klamath Falls and Eugene City. Pioneer Elijah Bristow and others claimed the miners' trail went into the Calapooya Mountains by way of Coal Creek, a tributary to the Middle Fork of the Willamette River. An Indian woman who had been taken to the two Frenchmen's camp fled and came out to the Military Road near Corral

Springs, making these springs a clue to the location of the mine from which the Frenchmen took out $50,000 in one year.

Another cabin was found and lost again by two hunters on Steamboat Mountain in the upper Applegate River country, in Jackson County. Still another cabin, where stagecoach loot is buried, lies in the Cascade Mountains, possibly in the Hood River area. Last but not least is Old Bohemia's Mystery Mine, at the headwaters of Steamboat Creek, tributary to the North Umpqua – a mine that some people said was a fake and others thought was a rich pocket.

The Stories of A. A. Reiner

My reference to an Oakridge, Oregon, packer and guide who had taken parties into the mountains in search of the Two Frenchmen Mine brought a letter signed by the man himself. He wrote that he "had an opportunity to observe all the prospectors who came looking for the Frenchmen's Mine" and had a lot of stories to tell, and he added in passing that I had misspelled his name – correctly it is A. A. Reiner.

Al Reiner is now retired and living in Stayton, Oregon, with his wife Alice. In the spring of 1967 I was able to visit him and to take down the promised stories. I found him an odd and fascinating combination. He was a Portland, Oregon, furrier during the early part of his life and he returned to that same business in later years, yet he spent the intervening time living a forest idyll.

Finding the work of a furrier confining and having a passion for the out-of-doors, he set out from Portland in 1916 on a canoe trip up the Middle Fork of the Willamette River, debarking at Oakridge. At that early time Oakridge was a settlement of only thirty-five people, all travel was by horse and trail, and the tall timber of the Oregon Cascades stood virtually undisturbed. Al Reiner was twenty-eight years old and a confirmed city dweller, yet he made a decision to stay in the wilds of the upper Willamette country. Which he did – for fifteen years. Ten of these years were spent as a packer for the U. S. Forest Service, the other five as a private packer and guide.

From meticulous city craftsman to outdoor expert is certainly a startling change. But to know Al Reiner even when he was approaching age eighty as when I interviewed him is to realize that in his prime he must have been capable of almost anything.

A diary he kept during his years in the upper Willamette country reveals that in a ten-year period he traveled 25,000 miles on horseback, or the distance around the world, all at two-and-a-half miles an hour; and the quality of his nostalgia suggests that life ought to move at no greater pace and always along dim forest trails, and he talks of the horses and mules which served him with an affection that other men give to old and devoted friends.

On up the Middle Fork about twenty-two miles south and a little east of Oakridge, Reiner built himself a cabin at the mouth of Coal Creek. This was at Campers Flat where many treasure hunters went by his door following the route popularly believed to be the one taken by the mysterious French miners when they went into the mountains. The man who declared this to be the right trail was the pioneer Hunsaker, the last person alive who had personal contact with the two Frenchmen.

Reiner remembers Hunsaker as a tall, heavy old man, ill and seemingly poor. Once they camped together at a spring at Sand Prairie, where Hunsaker was taking the sulphurated water for his health. He had been a preacher and still had a wonderful speaking voice and told tales well. When the talk turned to the Two Frenchmen Mine, Al Reiner listened with avid interest.

On returning from trips to Eugene City the French miners had frequently stopped at his farm at Pleasant Hill to get their mules shod, Hunsaker said. On leaving his place they went through Oakridge and on to Campers Flat. From there they took the Bristow Trail which follows the ridge between Coal Creek and Indian Creek, about six miles up to Bristow Meadows (or Prairie) in the high country. But where they went from there Hunsaker admitted he did not know.

Most of the people passing Reiner's cabin and toiling on up the steep Bristow Trail were making the pilgrimage to try to ferret out the miners' old trail and follow it up to the famed Two Frenchmen Mine.

One summer a man named Morton hired Reiner to pack a

pile of supplies up to Bristow Meadows for him. Morton was from Grays Harbor, a tall man about fifty-five years of age who wore a Luger on one hip and a Mauser on the other.

Morton was an experienced woodsman but he had with him an English couple who were the veriest of novices in the wilds. Frequently resting as they climbed the Bristow Trail they were barely able to stumble on to a cabin Reiner knew of. Even its shelter was a dubious blessing as mice routed the couple screaming in the middle of the night. The following day the Englishman built himself and his wife an outdoor bed, elevated on posts, but it crashed down in the night.

Morton had a map (or claimed to – Reiner says he never saw any such map) which would lead him and his companions to the lost mine. But a fear of the high country and the magnitude of the mountains made the woman too uneasy to go prospecting and too uneasy to stay in camp alone. So her husband was forced to take her back to Oakridge. He then returned to pursue the elusive mine with Morton. Just how much prospecting the men did Reiner is not sure. For that summer Morton started building himself a cabin on a smaller meadow to the northwest called Ranger Prairie. It turned out that he had been a ships carpenter by trade – and a fine one.

But the summer's activity did not yield the mine and the next season Morton returned to the quest. This time he brought with him a Grays Harbor mailman, his wife and five small children. Again he hired Reiner for the packing job – a complicated one this time because three of the children were too tiny to negotiate the trail. How to get them up to Bristow Meadows? Reiner selected an ancient, patient packhorse, Old Bud, put one child in the saddle and two in canvas bags called *alforjas* (locally pronounced alfarkie), one to each side of the horse. In crossing the river the children started screaming, which caused Old Bud to stop in midstream and look back puzzled at each canvas bag as much as to say, "This is a strange noise for a pack to make," but he soon plodded forward and carried the children to the high country without mishap.

Al Reiner learned that the English couple the year before and the mailman this year had paid all the expenses for the trip. He began to see that Morton's alleged map of the Two

Frenchmen Mine netted him free summers in the mountains and companions to help him prosecute his search. Al wondered whom Morton would get to pay his expenses the third year.

But the next summer Morton appeared alone and seemingly broke and went up to the meadows by himself. Later on Reiner packed in some supplies for him and was amazed to see that the former ships carpenter had finished his Ranger Prairie cabin. It was a large cabin, about forty feet by twenty feet, with a large fireplace at one end. The interior was lined with tin to make it rat proof. All in all it was a beautiful piece of work.

From then on Morton hunted the lost mine alone. Once he even confided in Al Reiner that he had found it, and to prove his contention he showed what he claimed was a human leg bone. He also located a magnesium prospect a mile and a half from his Ranger Prairie cabin, where he built himself a second, smaller cabin.

Other people came and went, also hunting the lost mine. They came with all kinds of gadgets, rods and metal locators. Some sought the remains of a cabin but others looked for quite different evidence: a dry digging, a well-traveled trail, even an old pole barn. One party located some blackened stones and spent the summer digging around what they believed to be an old hearth of a fireplace; Reiner thinks they had found only the remains of an old campfire.

Early one spring an old man and his wife were set down with their gear at Campers Flat near Reiner's cabin. The woman, partly crippled, had refused to accompany her husband prospecting unless she could bring her rocking chair, the better to take her comfort while waiting in camp alone while he spent time out in the hills. So here she was, rocking chair and all.

That summer as Reiner went about his packing duties he noticed that little by little the man was working his camp, including his old lady in her rocking chair, up the trail to Bristow Meadows, prospecting as he went. By fall he had wife and rocking chair fourteen miles up into the high country.

Then Reiner was asked by the Forest Service to go up and

pack the old lady down as she had been taken seriously ill. Reiner had a hard and anxious time getting the sick woman down the trail on horseback. He was much afraid she would die but she made it down to Campers Flat, where a car awaited to take her to the hospital. A few weeks later Reiner saw the old man coming down the Bristow Trail pushing a rudely-made wheelbarrow on which was stacked all the couple's gear. Riding on top was the old lady's rocking chair.

Reiner never hunted the Two Frenchmen Mine himself, but from talking to Hunsaker, Morton and assorted treasure hunters over a fifteen year period he has evolved a theory of his own.

Most of the Ranger Prairie and Bristow Meadows area drains to the north and west into Steamboat Creek, a tributary to the North Umpqua River. Just north of Steamboat Creek headwaters lies the rich Bohemian Mining District. With binoculars one can see from the high meadows to the mine dumps on Champion Creek.

If the Frenchmen went up to Bristow Meadows and then followed Steamboat Creek back to the Bohemian mining country it was certainly the long way round, but such a ruse might have worked in eluding the curious, who would not have expected them to have doubled back north. So if Reiner's intuition in the matter is right, the lost mine lies not in the high meadow country, not along Steamboat Creek, but somewhere in the noted Bohemian Mining District from which so much wealth has come.

Some Old Cabins — Found

While Mr. and Mrs. K. C. Boedecker were living in Portland, Oregon, they used to go gold panning as a hobby. One summer they drove up the North Umpqua River and on up Steamboat Creek. About nine miles above Steamboat Falls and Park they crossed a concrete bridge spanning the creek. Only a few yards above the bridge and buried in the brush near the creek they found the remains of a small cabin. It was not more than six feet square and completely collapsed, its logs lying crisscross where they had fallen. Only the foundation was in place. Nothing indicated the exact age of the relic or its past history but the Boedeckers felt it must be at least a century old.

In doing some panning at the spot they found a little gold. Downstream they got only colors; upstream nothing. On some future trip they hope to look up county records to learn if the cabin was built on the site of some old mining claim.

Also, Norman O. Rosenberger, Sr., of Sandpoint, Idaho, has written of two curious discoveries made while he was living at Grants Pass, Oregon.

"You will find Merlin, Oregon, on the map about eight miles northwest of Grants Pass. In 1950 my son and I were deer hunting a little northeast of Merlin when we came upon an old logging camp. We went up an old log road and down the ridge and run into an old trail. We followed this up and around the ridge, and right off the trail we found a very old and small cabin with a tree growing through the floor. You had to be right next to the cabin to see it, as it was in trees and brush. There were a few chunks of quartz rock laying around. They were very white rocks but not knowing a thing about gold them days, never paid much attention to them. I am sure that within a few days I could find the place again."

Years before that while hunting deer in the Applegate country, Rosenberger found an old mine, "very well timbered with lots of tailings outside, no roads within two miles of it." About three-fourths of a mile down the hill he discovered a tree which was carved with a name and the date 1850. "I was back to the mine twice after that," Rosenberger says, "and I know I can find it again."

What of Corral Springs?

Since the name Corral Springs has often been mentioned as a clue to the location of the Two Frenchmen Mine, L. J. Edmonds of Cave Junction, Oregon, has written to ask its location.

On looking into the matter I began to understand the confusion in connection with these springs. For the Metzger *Atlas of Klamath County* reveals three places of that general name:

Within the Fremont National Forest, T26S, R11E, Section 16, lies one Corral Spring.

Near the John Smith Road within the old Klamath Indian Reservation, T31S, R9E, Section 23, lies another Corral Spring.

Within the Deschutes National Forest, T27S, R8E, Section 7, lies Corral Springs.

A study of the maps reveals that the last named would have to be the one meant. It is in the plural, Corral Springs, as in the Two Frenchmen Mine story, and is located about five miles northwest of Chemult and only two or three west of Highway No. 97. In this vicinity the old military road did in general follow the present route of Highway No. 97, branching off northeast into the mountains where Highway No. 58 now crosses the Cascades. Which means that the fleeing Indian woman might well have come across soldiers working on the military road near Corral Springs. But how far and in what direction had she traveled since leaving the French miners' camp? Without that knowledge the correct location of Corral Springs proves not such a helpful clue after all.

AUNTY PERRY AND THE PINT OF NUGGETS

LOST TILLAMOOK MINE (Hult, *Lost Mines,* pages 105-107).

When Indians of the Oregon Coast Range brought nuggets to traders in Gales City and Forest Grove, avaricious white men attempted to follow them back into the hills — in the area of what is now called the Tillamook Burn. The white men had no luck, but did surprise an Indian and his son coming out from the mine with a load of gold dust and murdered them.

The widow of the murdered man later drew a map of the mine location for a nephew. The map is supposed to have passed to white miners and to have spurred the search.

Sol Emerick, a West Tualatin Plains farmer who was friendly to the Indians, was told by an old crone that he would find the mine where a stream ran into a lake in a black canyon — but the description was too vague to be of help. Emerick spent much time looking for the mine but failed to find it. F. L. Watrous (the initials were wrongly given in *Lost Mines*) did discover an old digging he believed to be the source of the Indian gold but it was on private property and he could not file a claim.

The More Medicine the Better

Letter of comment from William M. Perry, McMinnville, Oregon:

"My grandmother, Elizabeth ('Aunty') Perry was a pioneer of 1845 and was for many years the only doctor between Portland and Astoria. She died when I was ten years old but I have a small boy's memory of a story regarding her, an old Indian and a pint of nuggets.

"She was living on the old donation land claim about three miles west of St. Helens, Oregon. The Indian was short and fat with a consumptive cough. He probably had tuberculosis and he came to grandmother for a cure. She presented him with a couple of eight-ounce bottles of cough syrup, telling him to take only one spoonful two or three times a day.

"This consultation took place on the back porch of the pioneer house. The Indian was not to be hurried. He sat rocking back and forth, saying something to the effect that if a little medicine was good, more was better. Grandmother rocked beside him and said, 'No, more will give you a big sickness in stomach.'

"The Indian had brought with him for payment, an old-fashioned green glass, pint Mason fruit jar full of gold nuggets. But grandmother refused to take so much gold; she would accept only two small pieces from the jar.

"She asked the Indian where he found the gold. He pointed to a notch in the Coast Range of mountains to the southwest and said, 'Three walks, two sleeps' – meaning it was a three-day walk away. Later the smoke from one of the Tillamook fires came up right behind that spot."

A man who has often gone in search of the Lost Tillamook, Don Viles of Garibaldi, Oregon, has evolved a theory about its possible location which he expounded in the October 1967 issue of his publication *The Buccaneer.* The clues are a bit more explicit in his version than in mine, the Indian supposedly having said, "You will find the yellow metal in a black canyon, by a blue lake, near a white cliff."

Don Vile writes: "There are very few lakes between Tillamook Bay and the valley land east of the Coast Range, but there is one that is, or at least used to be, very prominent: Skookum Lake.

"The very rugged Skookum Lake area could be the site of the unfound gold. The lake when I knew it was large, being created by a wild beaver company with big ideas, which had engineered the construction of a dam containing many thousands of board feet of beaver-cut poles. The lake lay in a bowl-like depression that old timers thought was the crater of a small, extinct volcano. And when you visited the area you soon noticed that the surrounding crags, Gold Peak, Edwards Butte and Grindstone Mountain aligned themselves so as to look like the ragged edge of some blowout of long ago.

"Limestone deposits, which do exist in the region, could account for the white cliffs mentioned. The black canyon is hard to figure unless you remember that in those days all this part of the country was covered with dark, cool forests whose heavy shade made for many a black canyon.

"As for the blue lake, a body of water the size of Skookum was large enough to hold back the encroaching forests and to reflect the blue of the sky.

"Skookum is an Indian word, and a well-worn Indian trail ran by the lake — a trail that was later used by settlers in the area.

"So there are many points in favor of considering the region around Skookum Lake as a possible location of the Lost Tillamook Mine.

"Now, we don't want to mix you up, but here is another thought. There is a small lake behind Lee's Camp on the Wilson River called Blue Lake. The white cliff and black canyon could be references to a waterfall at the head of a box canyon — and Cook Creek, a tributary to the Nehalem, is loaded with box canyons and waterfalls."

BIG CITY TREASURE MOST POPULAR OF ALL

PORTLAND TREASURE CHART (Hult, *Lost Mines,* pages 109-110)

When Judge Everett Smith died in Seattle in 1933 his son Irving D. Smith found a totally unexplained item among his effects. It was a treasure chart, made on a piece of tracing cloth six by eighteen inches, with the following legend in blue lettering:

PORTLAND OR. SIMS MONEY IS
BURYEED 5 FEET DEEP AT THE
TWO GRAVES
NORTH FROM BARN TEN FOOT
EAST
KEEP THIS CHART TILL GET WELL
ALL HAVE DIED
PORT
PIKE ROAD TOO MILES FROMROM
PORTLAND OR.
GRAVES $3000 $3000
TO FIND
SIMES DIED IN THE ARMY 1864
KEEP CHART
CHART OFF
Feb 1862

Interspersed in yellow lettering were the additional words:

FEB 14-1862–HOUSE–BARN–BIG STUMP–ORCHARD

To the left side of the chart was a crude drawing showing eleven buildings, two of which had spires, a pond, and a background of hills. From the smaller spired building on the left, a blue line ran to the right and upward clear across the chart. Along this line the words were repeated:

PIKE ROAD TOO MILES FROMROM PORTLAND OR.

Above the line was the sketch of a house; below of a barn; and near the barn two short heavy lines followed by the word GRAVES.

Big City Treasure Most Popular of All

So far Portland history has disclosed no citizen named Sims or Simes who died in the Civil War and no road called the Pike Road.

Out the Great Plank Road

The response to the story of the Portland Treasure Chart proves there is no more potent combination than that of an intriguing mystery involving a treasure buried within the confines of a large city. More questions, comments, suggestions and possible solutions in connection with this old cache have been received than with any of the other treasure tales told. Neither the story of the romantic Neahkahnie Treasure Chest nor that of the famed Blue Bucket Mine has managed to stir up so much activity.

Certainly the Portland Treasure Chart is so cryptic and yet so definite of facts that it holds out a promise of decipherment if only the right key can be found. And its location, not far from downtown Portland, gives city dwellers a chance to stay home and yet join the treasure game; it also provides a convenient base of operations for out-of-town treasure hunters. So it is understandable that this particular chase has been a popular one.

While I was still living in Portland Mrs. Arthur House telephoned to say she was glad I had merely presented the facts about the treasure chart and had not tried to solve the enigma myself. This had left her with a challenge she enjoyed, and she had embarked on an attempted solution. Later in a letter she put down her deductions and tentative conclusions:

"In describing the chart you stated the blue line marked Pike Road started from a small spired building on the left and ran diagonally up to the right. This building may have been an early Portland church or may simply have been meant to indicate a starting point in downtown Portland. The Oregon Historical Society gave me the locations of several churches in the 1850-1865 era, all down on Second, Fourth and Fifth Avenues.

"According to conventional map directions, the Pike Road runs from town, shown in the southwest, up to the northeast. But the person who drew the map may not have had these formal designations in mind, so possibly one must turn the map to orient it correctly. This was suggested to me by the words: 'Above the line was the sketch of a house; below of a barn, and near the barn two short heavy lines followed by the word *graves.*' On the chart the words are, 'Two graves *north* from *barn* ten foot east.' North thus seems to be below the line and toward the bottom of the chart.

"If the map is turned to the proper orientation, then the pond could have been Goose Hollow Lake which used to lie where Multnomah Stadium is now located. My husband remembers this as a lake formed by drainage from the West Hills, which could be the background of hills indicated.

"Starting with the thought that main roads in early days were called pikes, I talked to an old pioneer of Forest Grove, who questioned other old timers. They all said that the road from Portland out that way used to be called the Pike Road. This is what is called Canyon Road now.

"My husband and I picked a starting place at Third and Jefferson and headed out Canyon Road. When we reached 'too miles fromrom Portland' we came to where the bulldozers were slicing up the hills for the driveways into the new Zoological Gardens. That area before it became the Zoo was a golf course, and long before that the site of the poorhouse.

"I did not find any record of Sims or Simes. I think one of these spellings is a mistake due to the soldier's illiteracy."

Ronald L. Gluth of Portland also wrote that Canyon Road "was a wood plank road and called the Pike," and he said he had discovered an old photograph of Portland showing "several tower buildings as on the treasure map." He was planning a hunt for the cache until the Vietnam war crisis caused him to enlist in the Air Force. Now his treasure hunting will have to wait until he has become a civilian again.

If Canyon Road is the Pike Road referred to, from what point should one measure the two miles? Interestingly enough there is a plaque near the center walk in the South Park Blocks in Portland, at the corner of S. W. Park Avenue and S. W. Jefferson, which reads:

THE GREAT PLANK ROAD

First planks for the Canyon Road connecting Portland with the rich Tualatin Valley were laid near this spot in September, 1851. As a community effort it brought farm produce to our docks and established Portland as a market for the Old Oregon Country.

Lang Syne Society

1960

Yet what is called Canyon Road today starts at the Vista Bridge, where Jefferson turns to ascend the hill.

If you believe that Canyon Road is really the Pike Road meant you will have to pick your own starting point — and you may come to the end of the "too miles" at a more propitious spot than the bulldozed terrain found by Mr. and Mrs. Arthur House.

A Family Treasure Hunt

Still, there is the possibility that a completely different road was meant. Or so argues the spokesman for a Portland family of six who also started out on the trail of the treasure chart.

Ruth Burgess, mother of the family, writes:

"I read your book cover to cover one night. . . . and next day we decided to make a game of solving this one mystery, but the further we got the less it got to be a game and pretty soon excitement prevailed. . . . We all tried to figure out the word 'fromrom.' The only explanation I could think of was that 'rom' was an abbreviation put too close to the word 'from'. We called the Historical Society to learn if a Roman Catholic church existed in Portland around 1864. We were told yes and given the location.

"Our son Richard looked up 'pike' in the dictionary and found it could mean a mountain road. The nearest mountain was Mount Tabor and we discovered that a certain road which goes over Mount Tabor begins near the site of the old church. My husband Robert figured mileage from the church up that road, and just a little more than two miles away we discovered a cemetery. We felt this couldn't be just coincidence.

"We invaded the cemetery one Saturday afternoon. We

looked for dates around 1864, and the names Sims, Simes or Graves (a possibility). We searched for two hours without success. Then our son Robert said that maybe Sims or Simes was just a nickname like the name 'Butch' that he was outgrowing. We thought this a good suggestion, and started out again, this time looking for names that were similar.

"About then, Julie, our daughter, yelled that she had found a house that looked as if it had been a barn. We thought so too, as it had all the marks, including a square roof, and seemed quite old. This was along a street on one edge of the cemetery.

"Not long after that, when we had worked ourselves over to the other side of the cemetery, our oldest daughter Sherry called to us excitedly. She had found two gravestones together – one stone tipped and much smaller than the other.

The graves were those of two children whose family name was satisfyingly similar to the ones they sought. One stone was undated. The other showed the date 1863, below which was carved "No More"—which the Burgess family interpreted as meaning the same as "All Have Died."

"We were so sure these were the right graves we looked no farther," Mrs. Burgess continues in her letter. She thought the stumps referred to might be artificial stumps erected by the Woodmen of the World as gravestones, of which there were a number in the cemetery. She also speculated that it might have been the father of the children who made two caches of $3000 each and who drew the treasure chart before he died in the Civil War. But she could not fit in all the dates. And of course the Burgess family did no experimental digging since they are lawabiding citizens and would not think of desecrating a graveyard.

(The above account does not reveal the name of the road the treasure hunting family followed, the name of the cemetery arrived at, or the surname found on the two headstones. These are omitted in an attempt to forestall any such vandalism as occurred in "The Case of the Ruined Graveyard" [*Lost Mines,* pages 130-131]. I warn readers who may follow the clues and locate the graves that the jail sentence for disturbing this cemetery will be a long one—at hard labor, I hope.)

Mrs. Burgess said they were thrilled to figure out something no one else had thought of, and she concluded her letter by asking what I thought of their clues. I answered that at the first opportunity I would visit the site of their discoveries and inspect the evidence.

When I was able to make this excursion I found the old cemetery a quiet and peaceful oasis, wonderfully tree-shaded, with traffic flowing by so mechanically on the busy city streets which hemmed it in that the green lawns seemed lone and isolated.

The gravestones in question marked the burial place of an "Infant Son" and of another boy who had lived but "2 yrs 2 ms 26 ds," born of the same parents. I noted one thing Mrs. Burgess had not mentioned: The family name (which I will not identify) is spelled differently on the two headstones. My own feeling is that this has no bearing on the Sims and Simes spellings but represents an error made by an early stone carver.

The Burgess theory seems to me a remarkable piece of sleuthing. But like Mrs. Burgess I cannot fit in all the dates. If Sim(e)s is the man who died in 1864, then the date 1862, twice repeated, would seem to be the year he "buryeed" the loot "5 Feet Deep At The / Two Graves." Yet the gravestone dated 1863 would not have been in place at that time.

While the house that sharp-eyed Julie Burgess spotted does look as though it might have been a barn, I felt that it stood too far away from the graves to qualify as the landmark mentioned.

Another difficulty: Research into early-day records has revealed that the father of the children buried under the stones did not die in the Civil War but was still alive as late as 1871.

So we are forced to leave it at that, the mystery not quite solved. But the adventures of the Burgess family give us a charming picture of treasure hunting as it should be, involving fun, excitement, teamwork, brainwork, exercise, fresh air, a new interest in history and a decent respect for property.

When next this formidable treasure hunting family takes to the field, surely "track's end" will be in ground where they may dig. My own fond vision is of the dirt flying, the prize within their grasp . . . .

And Where is the Treasure Chart?

A number of people have written to ask: Where is the treasure chart? Who owns it? Could they get permission to see it? They feel that a study of the original might yield information that no mere description can give.

The source of my brief account was a note headed "Treasure Chart," published in the *Oregon Historical Quarterly,* Volume 41, Page 109. My assumption was that Irving D. Smith, heir to his father's effects, had donated the treasure map to the Oregon Historical Society, the news item in the *Quarterly* being written at the time of acquisition. But those who have asked at the Society's headquarters in Portland have been turned away. Recently an OHS staff member wrote me: "As far as I know or we could ever find out, the Historical Society does not have the chart." According to what I can learn Irving D. Smith does not have it either. It is lost somewhere between chairs. Hopefully, it may turn up yet.

A CORRECTED TREASURE STORY — AND MORE URANIUM LORE

IN THE LOST FOREST (Hult, *Lost Mines,* pages 112-113).

About forty miles north of Silver Lake in Lake County, Oregon, lies an isolated ponderosa pine forest, partly inundated with sand dunes. A cowboy hunting lost cattle paused at Sand Springs where he picked up a rock with a dull golden color. He later gave it to Judge Daly of Lakeview to be assayed. The rock turned out to be fifty percent gold, but by the time the report was returned the cowboy had strayed and could not be found. When Judge Daly died he left a provision in his will that the gold and the assay report should be turned over to the cowboy if he ever showed up.

This and other stories have given rise to a belief that a rich gold strike would someday be made in Lake County. Instead, in 1955 a discovery of autinite, a secondary uranium mineral, was made north of Lakeview. Men going about prospecting with Geiger counters should also keep an eye out for the cowboy's lost gold prospect.

Two Stories "Mashed" to One

Letter of comment from Leslie Shaw, Editor, *Lake County Examiner,* Lakeview, Oregon:

"John Roush, one of the discoverers of the White King uranium claims . . . . brought me your *Lost Mines* book. I first read the story of the Lost Forest mine because it was about Lake County, and the Blue Bucket story because there are some who believe the location to be in this county.

"I was especially interested in the Lost Forest story, as I had never heard it before. I have asked several old timers and they had not heard it either. Your story referred to Dr. Bernard Daly and said the doctor's will provided for giving the gold from the sample and the assay report to the cowboy.

"I know you are anxious for accuracy, so I will tell you I have checked and find there is no such reference in Dr. Daly's will. Your sources indicate the story came from an interview with Hal Ogle of Klamath Falls, who is an excellent authority on the history of this area but I'm afraid his recollection is in error this time. To hear Mr. Ogle speak about the early days, as I have, is a notable experience."

Leslie Shaw kept this discrepancy in the Lost Forest story in mind, and at a much later date in going through the old files of his newspaper happened upon an item which seemed to explain the error about Dr. Daly — and even to reveal the origins of the legend itself. He mailed me a duplicated copy.

The news story, published in the Lakeview *Examiner* March 12, 1925, under the headline "GOLD DISCOVERY REVIVES LEGEND," tells of a recent strike near Paisley which had created a gold rush. It had resulted from the discovery of two men, G. S. Taylor and G. Picklesimer, of a rich quartz ledge. The ardor of some of the stampeders was cooling but the original discoverers claimed that their continued prospecting of the ledge made it look even better than at first, and they were waiting breathlessly for their assay returns. They had found pieces of pioneer wagons and discarded equipment in the vicinity which led them to believe their find might be the Blue Bucket Mine.

And in reporting reactions in Klamath Falls to the Paisley strike, the *Examiner* quoted the following from the Klamath Falls *Tribune*:

"More fuel for the fire of gold seekers lies in a tale told by a local physician. A cow puncher, cantering over range country near Paisley, several years ago, was attracted by a certain black rock. The cow puncher clipped off a piece of the rock and was amazed at its weight. He shoved the piece of rock into his pocket and promptly forgot all about it. Some time later he rediscovered the piece of rock in his pocket and bethinking himself of Dr. C. P. Mason, local dentist whose hobby is metallurgy, took it to him. Several weeks later Dr. Mason called the cow puncher up to his office and told him that the piece of rock assayed gold over $5000 to the ton. But by that time the cow puncher could not remember where he had run onto the rock except that it was somewhere in Lake county and probably in the northern part.

"Apropos of the Paisley gold strike, is an incident related by J. A. Gordon, president of the First National Bank of Klamath Falls. About a year ago a man came into the bank with a small bottle full of what appeared to be light colored sand. 'What is it?' asked Mr. Gordon. The man, whose name was withheld, answered, 'Gold. Will you send it away to be assayed? In a day's walk from here, I can arrive at a place where I can get some more.' To make a long story short, the alleged gold was sent away to be assayed. The government assaying office found the small bottle of sand to be practically pure gold. In fact, it was found to be 847 and a half thousands pure. Now the government wants to locate the man who sent the gold and make inquiries as to where the metal was located, but according to last reports their efforts have been unsuccessful."

Carefully note how certain details from each of these two incidents dovetail with elements of the Lost Forest story. The story about the cowboy and his rock fits the Lost Forest story —

except for the rock's black color; and the cowboy did leave his sample with a man of substance in Lakeview for assaying, but somehow Dr. Daly's name has become substituted for Dr. Mason's. And the tale of the man who left the bottle of yellow sand with the Klamath Falls banker and disappeared before the assay report came back would seem to be the origin of the ending of the Lost Forest story in which the cowboy disappears before the richness of the assay is known.

What happened, it becomes plain, is that the two stories, in being handed down by word of mouth, became fused into one — and it was a later "mashed" version which Hal Ogle heard and I wrote down.

Strangely, neither of the two episodes in the *Tribune* story has any direct connection with the Lost Forest area. One relates to Klamath Falls and the near vicinity. The other is somewhat confused as to locality, for the item first says the cow puncher was riding the range near Paisley when he picked up the black rock but later says he could remember only that he picked it up somewhere in northern Lake County. Since Paisley is in the center of the county, and definitely not in the north, this seems a contradiction.

All of which is a perfect illustration of why treasure hunting is so wild, precarious, misleading, maddening, stimulating, aggravating, suspenseful and fun.

Saga of a Uranium Prospector

In giving only a passing mention to the uranium mines of Lake County I find I missed a good story. Indeed it is such an amazing, exciting story it should prove an inspiration to all newly groping prospectors and treasure hunters. For the original Lake County uranium discovery, as well as much of its subsequent history, was made by an amateur prospector who didn't even own a Geiger counter.

Don C. Tracy, a butcher and meat cutter who had operated his own business in Lakeview, Oregon, knew absolutely nothing about mineralogy or geology until he was about forty years old. Then in 1953 he found some beautiful, many-hued, semi-precious stones no one could seem to identify. This so frustrated him he borrowed books on gem stones from the local library and was led into a home study of mineralogy and geology, sending for advanced texts to the State Library. He, his wife Irma, and their four boys, Brent, Gordon, Sid and Greg, began making weekend trips into the surrounding country to gather stones, just for the fun of it. Tracy put together a rock polishing machine and built up a gem collection of which the whole family was proud.

And what of those gem stones he set out to identify? They proved so rare they were named Tracinite.

In the 1950s uranium was the big thing – a word as exciting as the magical word *gold* in early gold rush times. According to the experts, Oregon was not a state expected to yield any uranium. But Tracy, getting the urge to apply some of his newly won book knowledge, went prospecting for uranium in the Lakeview area anyhow. This was in 1954. He was encouraged in this folly by his wife's uncle, John Roush, a longtime prospector. Roush said he could see no reason why there shouldn't be uranium somewhere in Lake County.

Tracy had almost immediate success. Only twelve miles from Lakeview on Thomas Creek in the Fremont National Forest he discovered rocks showing crystals of such promising nature he immediately staked some claims he named the Lucky Day. But having absolutely no equipment he had to borrow a dark light from a friend to be sure he was right. Yes, the crystals did fluoresce the color of uranium oxide. Encouraged, he sent a sample to the Oregon State Department of Geology, which reported traces of autinite, the crystal oxide of uranium, and urged him to keep looking.

By then winter had closed down and the spring of 1955 was so wet and muddy it was late in May before Tracy could get back out into the Fremont National Forest. Again he had quick success. On Augur Creek, the next stream east from the locality of his first discovery and still only fourteen miles from Lakeview, he picked some rocks out of a hole left by a tree which had uprooted during the winter. The rocks appeared to be full of the crystal oxide of uranium, so Tracy dropped them off at Rouch's place for him to check with his Geiger counter. That evening Roush phoned to say the rocks were "hot" and Tracy better go right back and stake the area. Tracy had to work the next day, so the best thing he could do was meet Roush at five. Roush brought his Geiger counter with him and together they drove to the site.

It was decided that Roush, in return for using his caterpillar tractor for digging prospect holes and helping to stake claims, should receive a one-third interest in the mine; the remaining third was offered to Walter Leehman, Jr., who owned adjoining property which was "hot". Leehman accepted and put forty acres of meadowland into the project. John Roush later divided his third with his son Wayland Roush; Walter Leehman, Jr., divided his with his father, sister and brother.

The three original partners staked sixteen claims they named the White King Group. This is still the richest uranium mine in Oregon.

Earlier that spring Tracy's mother Josephine and his stepfather Bill Orr had come on a visit from Vallejo, California, and as his mother was much interested in minerals Don took her uranium prospecting as a treat. They drove up a road to within a mile or so of the spot where a week later Tracy was to make his White King discovery. They picked up some likely looking rocks but as they had no Geiger counter or fluorescent light to check them by, Don's mother took the rocks home with her, planning to have them assayed in California. Unfortunately the rocks were accidentally thrown away en route — and came to represent a notable lost opportunity.

For while Tracy, Roush and Leehman were still in the midst of staking their White King claims and trying to keep their activities secret, a party of men visited them. These men had been prospecting for uranium by plane but had been about to give up when rumors of Tracy's strike reached them. They had been able to scout out the White King site by air — and here they stood. The secret was out.

The new party, composed of Don Lindsey, Robert S. Adams, Clair Smith and Choc Shelton, prospected out from the White King site, and a mile away, right where Tracy and his mother had picked up the lost rocks, made a strike and staked five claims, known as the Lucky Lass Group or Mine.

Tracy's own discovery was so important that within three

months he and his partners had leased the White King properties to a big-time mining company.

The original lease was made to Dr. Garth Thornburg and Vance Thornburg, brothers who had been operating mines in several western states. They brought in as partners the Richardson-Bass interests of Fort Worth, Texas, and the Murchison Trusts of Dallas, Texas, to form the Lakeview Mining Company with Dr. Thornburg as president and James F. Poulos as general manager. Core drilling and underground shaft work went on during 1956 and 1957. In November 1957 the Lakeview Mining Company signed a contract with the AEC for the production of yellowcake (U308). It then built its own reduction plant a little north of Lakeview.

Following his discovery of the White King Mine Don Tracy abandoned the meat business and has since devoted his time to prospecting and taking care of his claims. Operating sometimes alone, sometimes with partners, he has built an impressive record of success.

He and two other men prospected and staked an old cinnabar mine, now leased to a mining company.

He acquired a cinnabar property seventy miles south of Lakeview near Likely, California — also leased.

He staked an old gold mine near Plush, Oregon. (Considering Mrs. Pierce's theory that puts the Blue Bucket Mine not far north of Plush, Tracy's discovery of a gold mine in the area is noteworthy.)

He was one of a group of men who prospected and staked eighty-three claims showing zinc, lead, silver, copper and gold, in an abandoned mining district near Paisley, which had flourished around the turn of the century and still showed old diggings, mine shafts and even old cabins. These claims were leased to the Hanna Nickel Mining Company.

In 1966 Tracy staked forty more scattered uranium claims, which he placed under contract to Western Nuclear, Inc.

In 1967 he staked forty anomaly claims — a term which means that though the ground shows a radioactive count, producing ore can be located only through drilling and development. Western Nuclear shows some interest in these claims.

But one must not picture Don Tracy as an old time prospector wandering through the country with burro and pack. He uses a jeep or other vehicle of four-wheel drive, or even an airplane when convenient, and he is now the proud owner of a Geiger counter, scintillator and other scientific gear.

Locating and staking are only the first steps in his work and he tries to stake only claims which will *move,* as he has discovered that holding property is too expensive.

Having staked a claim he must develop it to the point where he has something to show. Often he must show it to several mining companies before one is favorably impressed. Then he must deal skillfully, working through lawyers, to negotiate a favorable contract. If the leased property reverts to him he must go through the process again until he finds a company which will keep the mine in production.

Over the years he has discovered that mining runs in cycles. In 1959 the Lakeview Mining Company dropped its lease on the White King, no uranium was mined, and the reduction plant closed. After that the plant changed hands three or four times, either for purposes of promotion or in anticipation of future production. So for a few years mining in Lake County came to a virtual standstill. But in the middle and late sixties, no doubt due to the Vietnam war, it was *go* for all minerals.

Tracy agrees that his case should prove encouraging to other rank amateurs starting from scratch. "As a hobby I can't think of one more exciting and challenging than prospecting," he writes. "I like the adventure of getting out into the great outdoors, the hills, mountains, valleys and even the desert country. I try not to pass up anything. I have found old mining cabins, beautiful sun-colored bottles, even abandoned antique furniture, a gun, gem stones, arrowheads and minerals. Of course it isn't all easy. Often I come home exhausted.

"I don't think anyone should give all his time to prospecting unless he is already retired or until he finds a mineral deposit that has started bringing in an income."

He believes the outlook for uranium is very bright. For it is one of the minerals of the 21st Century. And things are different now than in the fifties when the AEC was the sole purchaser. The big companies, like General Electric, Westinghouse, etc., have lately been buying with an eye to the future; important mining companies are drilling and blocking out uranium mines for reserves.

A Nucliometer Pays Off

An account written for me by Clair D. Smith of Lakeview about the discovery of the Lucky Lass Mine agrees in the main with the statements of Don Tracy but differs on a few points and adds some interesting sidelights.

Smith first became intrigued with the idea of uranium prospecting in 1954 and was able to impart his enthusiasm to his brother-in-law, Don Lindsey, and their friends Bob Adams and Choc Shelton. Together they purchased a nucliometer, a machine supposedly sensitive enough to locate radioactive prospects from the air. In Bob Adams' Piper Super Cub they did aerial prospecting over parts of southern Oregon, northern California and Nevada, but in the spring of 1955 gave up prospecting from the air and took to traveling by jeep and foot instead. They had still not made a strike of commercial importance when they heard that someone had located an uranium body of considerable size in Lake County.

Through the report of a friend who saw pickup truckloads of rock coming out of the Fremont National Forest they traced the uranium mining activity, and one evening while riding around with the nucliometer in their jeep were able to locate the outcropping which Tracy and friends had staked. "We got a reading clear off the high scale of the nucliometer," Smith writes. "It was our first contact with ore in place."

The next afternoon, July 12, 1955, they went into the area again and started driving up and down old logging roads. Not far northwest of the Tracy discovery they stopped to inspect a logging road survey stake and saw the machine go completely off the prospect scale. "We switched to the next highest scale," Smith explains, "and got out of the jeep, walking around with the machine. At the top of the cutbank we got a still higher reading. . . ."

They dug some foot-deep holes in which they placed the machine. They found it "went off one-half of the high scale. This was the highest reading we had seen of anything we had discovered ourselves." At one point they left the vicinity and explored other roads but finding nothing else even faintly "warm" they soon returned to the "hot" ground.

They dug yet another hole, from whose depth they got the same high reading. Then, starting back to the jeep they happened to detour around a tree and saw the machine register a still higher reading. Hastily they scooped out another hole, set the machine into it – "and lo and behold it snapped the high scale of the dial. This was it! We were sure that we had found an ore body. . . . What to do! We had never staked a claim and really didn't know how. . . ."

But they were cagy enough to fill in the holes they had dug and pile brush over them, making the place look natural before they sped back to town. There they consulted Bob Adams who had not accompanied them on the trip. He promised to get the necessary equipment and go with them in the morning to stake the claims.

Clair D. Smith was operating a retail bakery in Lakeview which he could not leave in the morning hours. So his partners went to the ground ahead of him. He arrived about 1 p.m. in time to help finish running the lines and to drive stakes for a discovery claim and four claims around it. The word was already out and people were all over the place.

The next day they hired a front-end loader which on its third or fourth scoop uncovered ore green with uranium content. They later learned that the ore body peaked about a foot underground at the spot where the machine had snapped the high scale of the dial.

"Talk about excitement!" Smith writes. "The next day the discovery . . . was in newspapers, on the radio and TV all across the nation. People came from all over, some from 1000 miles or more away. . . . The first week after the discovery we estimated 2000 cars drove by in front of the open cut. Of the ore dug and piled by the side of the road, two or three tons must have been carried off piece by piece as souvenirs by sightseers. Our little town looked like gold rush days, with street hawkers on corners selling Geiger counters and scintillators. Somewhere between 7000 and 9000 claims were staked in our national forest that summer, though most of these have lapsed since then. There are still only two uranium mines in Oregon that have been producers, the White King and the Lucky Lass. The Lucky Lass is our mine, which we still hold."

With such entrepreneurs in the field as Don C. Tracy with or without mechanical aids, and Clair D. Smith and friends with their amazing nucliometer, it seems sure that other uranium strikes will be made and excitement reign again in Lake County.

THE DISAPPEARING DIARY

THE CAPTAINS' CACHES (Hult, *Lost Mines,* pages 133-146).

Captain James Johnson and Captain James Scarborough were both shipmasters and both early settlers on Baker's Bay, an indentation along the north shore of the Columbia River just inside its mouth.

Captain Johnson took a claim on the west side of the bay where the town of Ilwaco now stands. In 1850 he built a large, fine house, and a smaller one behind it for his Chinook Indian wife and their two sons. The captain served as a bar pilot and also as salvage master in connection with at least three shipwrecks. From one of the wrecks he recovered a fortune in $50 gold slugs. These were weighed by a neighbor, J. D. Holman, who gave their value at $20,000. Surviving evidence suggests that Captain Johnson buried his hoard in different places not far from his house, and then made daily walks of inspection to be sure his caches were still intact.

Captain James Scarborough, who also married a Chinook woman, retired in 1850 to a claim east of Baker's Bay on the eminence still know as Scarborough Hill or Head. He was the second recognized bar pilot and was also engaged in the fish business. He made money fast and is believed to have buried $60,000 in gold slugs somewhere about his property.

Both captains died so suddenly as to be unable to retrieve their gold: Captain Scarborough of an illness in February 1855 and Captain Johnson by drowning in the Columbia River about 1857. To this day their lost caches excite curiosity and bring a host of treasure hunters with metal detecting gadgets to Baker's Bay.

Exciting Document

An article, "Ilwaco's Gold Mystery," by Robert Nash which appeared in the December-January 1967 issue of *Frontier Times* is of interest, in part because of the accompanying illustrations. Pictures reproduced show the pioneer house built by Captain Johnson; remnants of the fireplace, all that remains of it today; and the tangle of trees and brush which have grown up along the path where the captain used to make his tours of inspection. Also included is a portrait picture of Isaac Whealdon and his wife, the couple who bought the place after the captain's death and whose children and descendants have been a chief source of information about the buried treasure.

Nash writes:

"After some research on the gold cache, I believe it's still buried. I was very fortunate in running down a lead on the whereabouts of a diary belonging to the late Mrs. Holman. This lady kept an accurate account of unusual events which took place in the early 1800s. She verified Mr. Holman's weighing of the gold slugs, explained that the captain salvaged gold from a ship, and recorded other events which took place in those years. The diary is now the prized possession of Mrs. J. Tobler."*

The idea that the Holman diary with entries concerning Captain Johnson and his gold hoard could still be extant was an exciting thought, making me eager to contact Mrs. Tobler and get permission to inspect the important document. A letter to Nash elicited the information that he no longer knows the whereabouts of the lady, she having left her last known address at Gold Beach, Oregon.

It is hoped that Mrs. Tobler, wherever she is, will make provisions for the preservation of the diary after her time. All too often the descendants of pioneers, being unaware of the worth of old letters, diaries and other memorabilia, throw out or destroy material of a priceless character. The diary of Mrs. Holman, irrespective of any light it might throw on the John-

* Quoted by permission of Robert Nash.

son cache, is a valuable document and should be placed for safekeeping in an historical museum.

Little Mary's Yellow Stone

A bit of history in connection with the Scarborough property has been sent by Charles R. Eastland of Astoria, Oregon:

"I am a descendant of the Roque Ducheney family who bought the donation claim of Captain James Scarborough after his death and lived on the hill for many years. My great grandmother was Mary Ducheney, the youngest of the six children of Roque Ducheney. As there were various relatives named Mary she was known as Little Mary. When she was only fifteen years old she met and married my great grandfather John Fitzpatrick.

"Fitzpatrick was born in Ireland in 1820. He was a Columbia River seiner, who had acquired rights to the seining grounds known as Fitzpatrick Sands. He was forty-five years old when he married Mary, and he sent his child-bride east to be educated in a Catholic academy. Not until she was age eighteen did they set up housekeeping in Astoria.

"Fitzpatrick made a lot of money off the seining grounds and invested it in real estate, including a big ranch at Skamokawa, Washington, and a fancy resort-hotel at Newport, Oregon, which he built. He was drowned in the big Columbia River freshet in 1894, after which Mary lost some of the property but hung on to the rest and managed it until she became old and ill. She then sold out and went to live with a daughter in California. She died in the early 1920s.

"There is a story in the family about something that happened when my great grandmother was very young and which she herself used to tell in later years.

"The Ducheney home was on Scarborough Hill a short distance from the old Scarborough home. One day Little Mary came in from play with a yellow stone in her hand. Her father Roque noticed it and took it from her. On examining it he found it to be a piece of gold. Apparently Little Mary had been playing with it for several days before she happened to bring it into the house. Where had she got it? 'Up on the hill,' was all her father could get out of her.

"This happened only a few years after the Ducheneys had purchased the donation claim. Possibly Little Mary had stumbled upon a piece of the captain's gold. Or perhaps somebody dropped the stone on the hill. Or there's even a chance there is a source of low-grade gold somewhere back on Scarborough Hill."

EVIDENCE: ONE GOLD SOVEREIGN

CHEVY CHASE TREASURE CHEST (Hult, *Lost Mines,* pages 153-157).

According to legend, a paymaster on a British Columbia railroad job absconded in 1864 with some $60,000 in gold sovereigns. He was believed to have placed the fortune in coins in a chest and to have hired an Indian canoe and its crew to paddle him from Victoria, B.C. across the Strait of Juan de Fuca and into Discovery Bay, where the Indians set him down on the beach with his burden. Not long afterwards a man believed to be the paymaster appeared at the cabin of a homesteader John Tukey and from him hired a horse to ride to Port Townsend. He was at that time unencumbered.

The absconder was arrested, tried and convicted of the crime. After serving his term he went to Texas and never returned to Discovery Bay to retrieve the treasure. Tukey and his step-daughter Mary Chase turned the homestead into the summer resort known as Chevy Chase Inn, and both continued to believe that the chest of sovereigns was buried somewhere about the premises. The mystery of the missing payroll has continued to fascinate, and Philip W. Bailey, the present owner of the charming old Inn, seems never to lack for treasure hunters, whom he allows on the property one party at a time, under contract.

In the Morning Mail

As Buck Richey of Radio Station KVI in Seattle, Washington,

was opening his mail one September morning in 1959, an old gold sovereign spilled from an envelope. The letter which accompanied it claimed to be written by an elderly Port Townsend woman who said that about a century earlier her father had befriended a man who had cleared out of Victoria, B.C., with a fortune in gold pieces and for his kindness had been rewarded with fifteen of the coins, of which the enclosed was one. The woman asked that if any publicity resulted from her revelation her name not be used but she failed to sign the letter — perhaps merely a lapse of memory.

No one on the KVI staff had any background for such a "revelation," so Richey put in a call to the Port Townsend *Leader*. whose editor, Richard F. McCurdy, related to him the locally well-known Chevy Chase treasure story, which matched perfectly the statements of the letter writer. The *Leader* editor, much interested to hear that a coin seeming to be one from the old hoard had come out of hiding, ran a news story on October 1, 1959, headlined: "TREASURE MYSTERY REVIVED. WOMAN CLAIMS GOLD SOVEREIGN IS ONE OF COINS LONG SOUGHT HERE."

Disappointingly, nothing further developed. The elderly woman who supposedly wrote the letter never revealed herself, but neither has any evidence been forthcoming to brand the thing a hoax. Although inquiries I have made for more information have met with no response it must be assumed that the coin was of sufficient age to qualify as one of the stolen sovereigns and the letter must have borne some evidence of authenticity.

Could the coin really have come from the paymaster's hoard? Or was the whole thing a practical joke by someone wanting to stir up some publicity?

A man I once met claimed to have done some research in British Columbia on this story and to have found absolutely nothing to substantiate it, but unfortunately I took no notes on the conversation and am now unsure how thorough his research may have been or exactly where it was done.

Recently I came across an early version of the story, one I had somehow missed before. It gives so much in the way of hard facts that the legend should be subject to verification – or, if the "hard facts" prove non-existent, to scotching.

This is the chapter "Treasure Trove" in James G. McCurdy's *By Juan de Fuca's Strait,* published in 1937. The author gathered his material mainly from what he calls "living epistles" – the people among whom he had a wide acquaintanceship in the Port Townsend area where his family were prominent pioneers. His account of the Chevy Chase Treasure brings us as close to the events as we shall ever get and where my version, gathered from secondary sources, differs from his I readily abdicate in his favor.

He gives the year as 1862 and states that John F. Tukey saw an Indian canoe with two paddlers bearing a white man and some kind of chest come into the inlet near the landing on his pioneer farm. The Indians helped the man carry the chest into the dense woods nearby, then returned to their canoe and paddled away. Some time elapsed before the white man appeared at Tukey's cabin, without the chest. He borrowed a horse and rode off to Port Townsend.

A few days later Tukey heard some exciting news from a mill hand who crossed his property.

McCurdy writes: "The mysterious stranger had reached Port Townsend just in time to catch the steamer *Eliza Anderson* on her up-Sound trip; but when Captain Fleming saw the passenger walk aboard he at once identified him as an absconding Canadian paymaster from Victoria. The captain had seen a poster with a likeness of the fugitive, and offering a reward for his capture. The poster had given the information that the man, who was a government inspector of roads, had disappeared with some six thousand gold sovereigns, or approximately thirty thousand dollars. The stranger had immediately retired to a stateroom and Captain Fleming as promptly turned his steamer about and retraced his course to Victoria where he delivered the fleeing paymaster to the officers of the law.

"Not long afterwards several Canadian officers arrived at Tukey's claim and interviewed him regarding the iron-bound chest that had contained the missing money. The officials had located several British Columbia Indians who had transported the paymaster to Dungeness Spit. Here he had paid and dismissed them and engaged two other Indians to bring him to Discovery Bay.

"None of these individuals was able to impart any useful knowledge, nor did they know what the chest contained or what became of it. The officials hired a number of men living on the bay to dig for several weeks along the beach and about the landing, but they were unable to locate the chest. The officers next traveled around the bay and offered a large reward for the empty chest, thinking that perhaps its contents had been removed and that it had been set adrift, but it was never found.

"Subsequently the paymaster was tried, convicted and sentenced to prison for a long term of years; but he could not be made to acknowledge his guilt or divulge what he had done with the stolen funds. As soon as released from prison he departed for Texas where he had a brother living and there he soon died."

A case of the proportions outlined by McCurdy, where a reward was offered and posters distributed, where a named captain turned his named steamer from its course to deliver a criminal back to the scene of the crime, where treasure digging and a new reward for an empty chest were authorized by government officials, and where a trial and conviction ensued — such a case would have been sensational in the extreme, and details of its unfolding development simply *must* have found their way into the newspapers of the day. Newspapers were being published in Port Townsend, Seattle, Olympia and Victoria, and if these events actually happened news stories should corroborate the details. This treasure story has flapped around in the legendary sphere long enough.

ANCIENT TREE BLAZE, STONE HUTS AND OTHER "SPANISH" MYSTERIES

LOST SPANISH SILVER MINE (Hult, *Lost Mines,* pages 171-177).

According to legend Spanish explorers in the 1600s or 1700s invaded the Pacific Northwest and discovered a vein of North Idaho's silver ore. They obtained large amounts of silver until their Indian slaves rebelled and killed all except one priest, who fled south. The priest drew a map of the mine location, which passed into the hands of young James Williams. He sought the mine along the St. Maries River, a part of the St. Joe drainage, and on the second creek which comes in from the east stumbled on a hole which gave access to a rich silver vein. He beat his way out of the mountains and started for Spokane to obtain backing to develop his bonanza. Near Rockford, Washington, he was arrested as a horse thief and though innocent was hanged. The vigilante who cut down his body found the treasure map in his pocket.

In connection with the remote possibility that early Spanish parties did penetrate farther north than reputable historians believe I would like to pass on a number of intriguing stories which have come to my attention. In the aggregate these stories certainly give food for thought, though admittedly they *prove* nothing. Or rather what they prove is that scientists and historians should give closer attention to such mysteries as they arise. In some cases commonplace explanations would no doubt result but in others evidence might be revealed indicating that persistent tales of Spanish activities in the northwest are based on truth. Who knows?

The Ancient Tamarack Blaze

In 1936 Leon A. Patten owned a forty-acre stump ranch in the northeast quarter of Section 29, T64N, R1E. This location is about fifteen miles north of Bonners Ferry, Idaho, not far from the Canadian border.

The area had been logged prior to 1925 by the A. C. White

Lumber Company, and Patten was now taking out buckskin tamarack trees which had been left standing — the term buckskin meaning the trees had been dead so long their outer bark had decayed and fallen away to reveal the smoother, tan-colored boles. Tamarack wood being unusually resistant to rot, Patten found the dead trees excellent for making cordwood. Working with his oldest son he felled a number of the trees, skidded them in to his woodyard and bucked them into four foot lengths. One day while splitting these lengths into cordwood, Patten chopped off the outside slab of a butt section and found, deep inside, an old tree blaze. The mark was shovel-shaped about eleven inches by twenty-one inches at its longest dimensions. A rough check of the slab which had split off showed an apparent age of about 200 years.

Patten, becoming excited, took the whole butt section home with him. He told a helper who was trucking the wood to town to stop by the Bonners Ferry Ranger Station and tell the Forest Service men working there about the discovery.

District Ranger John B. Thompson visited Patten's place, inspected the evidence and photographed the marked piece of wood and the cover piece, getting about a dozen prints in all. Two of Thompson's photographs are reproduced in this book: one of the center section of the tamarack butt showing the blaze; the other of the cover piece which had grown over the old mark.

In a report to his superior, the Forest Supervisor of the Kaniksu National Forest, District Ranger Thompson states that Patten had used a reading glass to check the ring count and had confirmed an age of about 200 years as represented in the piece that had grown over the blaze.

He then goes on to say, "Growth rings were very close and impossible to count without a good magnifying glass so only a casual count was made. However, I was able to check growth rings well enough to make certain that the blaze is well over 200 years old . . . . I firmly believe that this blaze was chopped into the tree over 200 years ago . . . . with a narrow-bladed, sharp, axe-like tool."

The blaze created a local stir. Patten put it on display, and, as he says, "We had lots of company for several months." He also wrote to the National Museum and other places to learn if his discovery had any special significance or value. Replies were inconclusive, and long after the excitement died down and the two pieces of tamarack butt had lain around for many months collecting dust, he sold them to George H. Unternahrer for the modest sum of $25.

Unternahrer was the manager of the Bonners Ferry J. C. Penney store and a man of many interests. He realized, as he says, that here was "a very valuable piece of Idaho history," and he was determined to save blaze and cover piece from oblivion. Using a magnifying glass he counted the rings in the split-off piece and found 176 years of growth after the blaze had been made. But he realized that the evidence must be studied scientifically before it could be said to prove anything. He made arrangements with the University of Idaho at Moscow to make such an examination, and in 1940 he delivered over to that institution the tamarack blaze and cover piece, as well as a green piece of tamarack cut from the same location to serve as a means of comparison.

Unfortunately, World War II intervened, with the University departments left so short staffed that the promised study could not be made. So in 1946 Unternahrer retrieved the blazed section and cover piece from the university and in 1949 presented them to the Idaho Historical Society in Boise, where for some years they were kept on display at the state museum. They are at present in the storage vaults of the society.

Leon Patten, now living in Spokane, Washington, first wrote me because he believes his discovery indicates that Spanish explorers passed through the Bonners Ferry area early in the 1700s. Considerable correspondence with him, with John B. Thompson and with George H. Unternahrer followed. All agree that, considering the angle of the cut, its depth and clarity, the blaze must have been made with a steel axe or axe-like tool. Unternahrer describes the axe used as "similar to a shingler's hatchet," which would also describe a Hudson's Bay axe.

How many years after the death of the tree before it would lose its bark to become the buckskin that Patten cut in 1936? Thompson estimates a minimum of thirty years. Add those thirty years to the 176 years of the ring count, then subtract the total from 1936, and the blaze must have been made in 1730 or earlier. Unternahrer, believing the tree had been dead sixty-five years, put the date of the blaze at 1695. Even allowing for inconsistencies in the tree ring evidence, it seems impossible that the blaze could have been made after 1750. And that is nearly sixty years before the first white man, the furtrader David Thompson, is known to have entered what is now the Idaho panhandle. True, there are echoes and legends, gathered by Dr. Claude E. Schaeffer, which indicate that Thompson sent two French-Canadian trappers among the Kutenai Indians of the Tobacco Plains country of eastern British Columbia as early as the winter of 1800-01; but seemingly these trappers had no contact with the Lower Kutenai Indians of the Bonners Ferry area. Besides, even this date is about fifty years later than the latest date of possible conjecture in connection with the blaze.

Of course Indians might have obtained white man's axes or other sharp tools through distant trading at an early date. If so, would this explain the blaze? Apparently not. Dr. Allan H. Smith, Chairman of the Department of Anthropology, Washington State University, Pullman, states:

"To the best of my knowledge, there are no reports in the anthropological literature of Indians blazing trees . . . . Indians were familiar enough with the terrain and their trails, and could follow trails with sufficient skill even if their path were unknown, to have made trail blazing an unnecessary luxury. And they were quite sharp enough to identify a locality with precision, even years later, without employing this technique. In short, from everything I know about Indians, particularly those in the Plateau area, I would doubt that the absence of mention of blazing technique in the professional literature is to be explained in terms of oversight. I believe it simply did not exist . . . ."

What then is the explanation? To date it remains an enigma. Leon Patten, the discoverer, John B. Thompson, district forest ranger of the Kaniksu National Forest, and George H. Unternahrer, merchant and history buff — all appear to be convincing, reliable witnesses. The story they tell seems worth exposure even though I am aware that the experts may dismiss their claims as nonsense. It is hoped that someone with special knowledge on the subject will have those tamarack pieces dug out of the Idaho Historical Society storage vaults (if they can still be identified) and make a scientific study which will settle the question once and for all.

Montezuma Treasure

According to accepted history the Spanish *conquistadores* in their explorations of interior North America never reached north of Colorado. They probably did not enter Utah – and yet . . . .

In 1922 all operations in the Utah desert town of Kanab came to a standstill while the entire populace surged out to a nearby mountain to help dig for hidden Montezuma treasure. It was not the Spaniards themselves, in this case, but Aztec carriers who were supposed to have fled with a part of the king's treasure – a hoard worth $10,000,000 – which they buried in the Utah desert, thereby thwarting their white conquerors. Or such was the tale told by prospector Fred Crystal who spent a decade of his life seeking clues to the location of the fabulous cache. He made two lengthy trips to Mexico, returning the first time with tracings of ancient Aztec petroglyphs whose duplicates he found on a rock wall in Johnson Canyon, twenty miles northeast of Kanab; and the second time with an old map he stated came from a Mexican monastery. The map led him to chalky White Mountain where he found a series of steps cut into the sandstone, leading one hundred feet up the face of the cliff. When preliminary digging revealed a tunnel entrance at the head of the steps everyone in Kanab turned out to help in the extensive excavations which followed. The town was deserted.

Sixty feet inside the tunnel searchers encountered a wall made from blocks of blue limestone, a material not available closer than thirty miles away. They encountered other obstructions and boobytraps but no treasure. After two years of fruitless digging the project was abandoned and the broken-hearted Crystal disappeared. But around Kanab the belief persists that somewhere inside White Mountain lies a part of Montezuma's storied treasure.

"Spanish Diggings" and "Spanish River"

The Spanish explorers probably never entered Wyoming, or Montana either – and yet . . . .

According to Struthers Burt in his popular book *POWDER RIVER Let 'er Buck*: ". . . . like the vague silver of the moonlight that haunts the Powder on August nights, the tradition of the Spaniard haunts the Big Horn and eastern Wyoming as it haunts so many other parts of the Northwest. No one knows just how far north the Spaniards got during the greatness of their seventeenth century. Too many records were destroyed when the Indian insurrection of the 1680s blazed along the New Mexican Spanish frontier. That the Spaniards got as far as Colorado we know. For a couple of decades at least, in the middle of the sixteen hundreds, Spanish pack trains laden with gold and furs were coming down from the north to the settlements near Santa Fe, and undoubtedly this continued up to the insurrection.

". . . . . we know that the earliest American trappers called the Green River in western Wyoming the 'Spanish River,' whatever their reasons may have been. Perhaps because the Green River ran straight down to the Spanish possessions in the Southwest.

"A hundred miles or so southeast of the Powder where Niobrara and Platte and Goshen counties come together, are the so-called 'Spanish Diggings,' huge abandoned quarries covering an area of ten by forty miles. Archeologists have decided, in contradiction to the name given these and the popular belief, that they are neolithic mine pits and factory sites. But there are other vague traditions.

"In 1865 members of General Connor's 'Powder River Expedition,' the first hostile military expedition to enter the Powder River country, reported the finding of the ruins of stone houses on the shores of Lake DeSmet at the eastern base of the Big Horns. A year later the supposed remains of a Spanish arrastra, a tool for crunching quartz, was found in the same vicinity. In 1874 Colonel Mills' expedition to the Big Horns also reported the ruins of stone houses. In 1866, miners coming out of the Rosebud country in Montana which is just north of the Big Horns, their hair flying back of them because of Indians, brought word of an ancient mining ditch they had come across. A ditch and traces of iron tools and of mining on a fairly large scale. But these remains, Iberian or not, have disappeared, all save the 'Spanish Diggings.' "*

The Stone Huts

The Spaniards would seem never to have entered Oregon — and yet. . . .

Struthers Burt in this same passage says that when "the French voyageurs and furtraders began to push west along the Canadian line in their search for the Pacific and Columbia, they constantly ran into relayed stories of 'tall men in armor, feathers on their heads,' tall men who had mysteriously appeared as far afield as eastern Oregon and had as mysteriously disappeared."*

And added to the reports of ruined stone huts in Wyoming's Big Horn Mountains are reports of stone houses in central Oregon.

Flora E. Bailey, whose statement about her uncle Dan Herren's picking up a gold nugget from a gopher hole along the route of the Meek wagon train is quoted in the story of the Blue Bucket Mine, goes on to describe the creek where the discovery was made:

"There was a group of small stone houses on a bench above the stream, sheltered by a still higher bank, and by trees. *My father, then ten years old, is the only one I ever heard who remembers the stone huts. He says they were round in shape and covered over the top and of different sizes,* averaging six or seven feet in diameter. Of the many expeditions sent to locate the spot, no one has ever given any account of the remarkable little huts." (The italics are Mrs. Bailey's.)

Yet there is one other tantalizing reference to these stone huts. Clark and Tiller, who in their book *Terrible Trail* give the complete text of Mrs. Bailey's statement, add this footnote: "Ben Houston, of Bend [Oregon] related that his brother Sumner Houston, of Prineville, says he found a group of stone huts

------

* Quoted by permission of Mrs. Struthers Burt.

on upper Silver Creek. In one of them was a newspaper clipping containing the Dan Herren information!"

But apparently no scientific examination or study of these stone huts either in Wyoming or Oregon was ever made.

Search for the Seven Cities

No, the Spaniards probably never extended their explorations into Oregon – and yet . . . .

In the mountains south of Hardman in Morrow County in 1931 a cow put her foot in a hole, revealing a shaft leading down into the earth. As later described in *Sundial,* a publication devoted to unusual discoveries in the fields of geology and archeology:

". . . . the tunnel descends into the earth at a slight incline, thru solid basalt, and strangely, no earth appears at the opening to indicate where the excavators had disposed of the debris removed in the digging of this tunnel.

"The walls have no appearance of blasting being done, but rather, they appear to have been chipped in a manner like that of work done by early day Indians, which leads one to believe it to be the work of a prehistoric race, rather than that of a miner.

"Another interesting feature in connection with this discovery is that at regular depths blockades or platforms were built in the tunnel. These platforms were built of materials foreign to that section and appear to have been held in place by wooden posts, now decomposed except the charred ends where the posts had apparently been burned to fit into the indentions in the rock wall. The . . . . workmen had penetrated into the tunnel to a depth of about 60 feet when they encountered another such platform, and here they abandoned their work."*

---

* Of interest because of its similarity is a shaft discovered by hunter Ed Mullard in the Jordan Meadows area of Vancouver Island, British Columbia, Canada. Mullard insisted that the shaft was neither a natural cave nor a mine. As described in the Victoria *Colonist,* November 11, 1960, "There were seven steps down, then an arch carved in the granite, and beyond that a rectangular gallery about ten feet long and high enough for a man to stand upright. In the right side of the gallery at its far end was another flight of seven steps down, and a second gallery. Here he found about a foot of water and he went no farther." The tunnel, whose location is now lost, is often referred to locally as a "Spanish Mine."

Treasure hunters felt the shaft might have been dug by an Irish sheepherder, Pat Shea, who according to local legend had stumbled upon a source of gold while following his flocks, and after a season of mining and hoarding had gone off to Portland to spend his accumulated golden wealth and had there drunk himself to death. Only when the tunnel failed to show the slightest trace of gold was the idea that this might be the old Pat Shea Mine relinquished.

The *Sundial* article expressed a hope that the mysterious tunnel might prove archeologically valuable in throwing light on regional prehistory but made no mention of a possible Spanish origin. It was an old timer I interviewed who in talking about the Morrow County tunnel told me a fantastic story which extended the search for the Seven Golden Cities of Cibola into central Oregon, connecting that search with the building of the tunnel. Nothing he mentioned tied the story to history, to fact, but since he was surely incapable of inventing such a delightful fabrication, the roots of the tale must be somewhere in Oregon folklore.

Rocky Bar Spanish Legend

The Spanish explorers almost certainly never entered Idaho—and yet. . . .

The discovery of the remains of old arrastras near Rocky Bar at the forks of Big and Little Elk creeks in the Boise River drainage of southern Idaho gave rise to a legend of such strength that the area is now known as Spanishtown. The Idaho Historical Society pooh-poohs the idea that the arrastras could have been built by Spaniards before the California gold rush, but a correspondent of mine says that her uncle who has done research into Idaho prehistory and has hunted the lost Spanish mine he believes lies in the Rocky Bar area "has a fit" over the Historical Society's interpretation. So the tradition dies hard.

And, as we have seen, in North Idaho there is the enigma of the old tamarack blaze and the legend of the silver mine James Williams is supposed to have rediscovered and to have lost again through death.

Art Jacobson's Lost Tunnel

At the Old Timers' Celebration in Harrison, Idaho, in 1968 a tall, dark, bronzed man in a black derby hat came up to greet me. He was Arthur A. Jacobson whom I had interviewed some years before while doing research on the Great Idaho Fire of 1910. He still lives on a ranch along the Coeur d'Alene River near Enaville, his home for many years.

As we ate our barbecued beans and ham in the Harrison City Park, Art told me that at one time he had become interested in the Lost Spanish Silver Mine and had made some calculations as to its most probable location. The Milwaukee branchline to Bovill and Elk River runs up the St. Maries Valley and when the location Art most favored for the mine turned out to be on railroad property he lost interest and didn't bother to go check out his theory. But later on when a friend of his, Julius Hagman, was at loose ends and wanted to do some prospecting Art said, "Why don't you give this a whirl?" and told him his ideas about the location of the lost mine tunnel.

In due time Hagman returned from the St. Maries Valley. He reported that to his surprise he had found a hole in the ground of the right description in the vicinity where Art thought it should be, with some decayed ropes hanging in it but so steep and eerie of descent he dared not enter it alone. Art promised to go back with him to help explore the shaft but before he could get away from the ranch his friend Julius became suddenly ill and died. Not sure he could locate the tunnel by himself Art has never gone back to look for it.

Where is the Map — or Copies Thereof?

A number of people, including James Hausam of Sedalia, Missouri, and Joe Rasmussen of Spangle, Washington, have written to ask what happened to the treasure map the vigilante found in one of James Williams' pockets after he was hanged. Unfortunately I have been unable to establish the name of the vigilante or trace the map in any way. That there was some vigilante activity around Rockford is true. Possibly old letters, reminiscences of old timers, local histories, or even newspaper files will yet yield the necessary references to pin this legendary story down to fact.

DIVING FOR LOST GALENA, SUNKEN BOATS AND OLD BOTTLES

THE SUNKEN ORE OF LAKE COUER d'ALENE (Hult, *Lost Mines,* pages 187-190).

The discovery of a rich lead-and-silver ledge in the Coeur d'Alene Mountains of North Idaho led to the development of that most fabulous of northwest mines, the Bunker Hill & Sullivan. Galena from the mine was shipped out of the mountains over a narrow-gauge railroad to the head of navigation on the Coeur d'Alene River and there loaded on barges to be taken downriver and across Lake Coeur d'Alene by steamboat. In the fall of 1889 the icebreaker *Kootenai* was pushing a loaded barge when, about halfway down the lake near McDonald's Point, the barge rode over on one side and toppled 135 tons of sacked ore into the water, at a depth of from sixty to one hundred feet.

Fred Wilson, fireman aboard the *Kootenai* that night, quit his job and attempted to salvage the ore with a grappling hook but could get nothing more than pieces of sacking. After World War II Wesley R. Johnson, an underwater salver, planned to use his Navy landing craft and other gear in an attempt to raise the ore but he died before he could make the experiment. So the sunken ore, valued at many thousands of dollars at today's prices, remains a fair prize for those who seek their thrills underwater.

Two Young Adventurers

In 1958 two young Army Air Force men, Spencer Campbell and Jerry McFeeley, became excited over this particular story and decided to have a go at the sunken treasure. Both were radar technicians, airplane pilots and skindivers — a wonderful combination for adventure. Stationed at Paine Field near Everett, Washington, they made trips to Lake Coeur d'Alene whenever they could get leave.

By then the only man alive who could give them personal advice was eighty-seven year old Ernie Weeks, a retired steamboat engineer. During his long career on the lake Weeks had served under Fred Wilson after he had risen to the position of steamboat captain. On their runs up and down the lake "Cap" Wilson had often pointed out to his engineer the place where the barge had dumped its cargo.

On interviewing Weeks, Campbell and McFeeley learned there had been a dock on shore opposite the spot where the ore had sunk. Remnants of the old dock would be a helpful landmark in knowing where to dive, but real difficulties lay in the way of discovering such evidence. Not only would the old pilings have rotted away but if any stubs remained they would be under water, because Lake Coeur d'Alene is now held at a higher level by the Washington Water Power Company dam at Post Falls.

Undaunted, the young airmen set out to explore a location to which Weeks had directed their attention. They made several skindiving descents and in about ten feet of water discovered some piling stubs they believed to be the underpinnings of the old dock. This proved so encouraging they also made some exploratory dives into deeper water.

"As we went down," Spencer Campbell said later, "the water which looks so clear and sparkling on the surface became cold, black and opaque. I was very much surprised at the lack of visibility at the lower depths. This seems to be caused by underwater currents bearing silt and mine wastes. We found the floor of the lake to be covered with a heavy layer of sludge, no doubt precipitated from these currents."

Not long after the expedition in which they made these dives the two young men looked me up in Seattle. We went out to a coffee shop and had a long talk, they telling in detail about their interviews with Weeks and their investigation of the site he suggested to them. They had decided that recovery of the ore would take more time than their Army Air Force leaves allowed them but they planned to continue the operation when their hitches were up in about a year. I have heard no more from them and imagine that when they returned to civilian life other responsibilities and duties captured them.

The Deep-Sea Divers

In the summer of 1960 there was interest in the sunken ore

from a new quarter. Jack Ross and Dick Greer, divers and treasure hunters, came out from St. Louis, Missouri, to assess the possibilities of using modern deep-sea diving and salvage equipment to raise Coeur d'Alene's sunken ore. Their arrival at Ozzie Walch's popular vacation resort, Land of Oz, on Carlin Bay, created a stir and their search was publicized in the newspapers, on radio and TV.

The men came armed with a copy of *Lost Mines and Treasures of the Pacific Northwest* and on arriving did some research of their own. In Wallace they went through old files of the *Miner* and in Kellogg consulted with officials of the Bunker Hill Company. They also learned that Marvin Jones, buff from a local history group, believed that the ore was sunk farther south in the lake than is usually believed, somewhere near Rockford Bay.

When at last Ross made a dive it was a little south of McDonald's Point, near a jutting of land called on my map Sand Point.

On surfacing Ross announced to reporters that he had found the "silver reef."

"It lies just above Sandy Point, in about a hundred feet of water – right where Ruby El Hult's book says it is – about 300 feet off shore." And he brought up a handful of material to prove it – "not galena ore, as many have said, but flour-fine concentrated lead, already milled."

In the end Ross and Greer claimed that the deposit on the lake floor was not so extensive as they had hoped and not worth the cost of bringing their deep-sea equipment from St. Louis and hiring men to help in the salvage operations.

Whether Ross, with so many eyes upon him, was embarrassed not to have found the mound of ore, or whether he had been misled by his own research is hard to say, but he was certainly mistaken about the nature of the treasure he was seeking. For the sunken ore could not possibly have been in the form of lead concentrate. In 1889, the year the ore sacks were lost from the barge, there was no milling of any kind in the Coeur d'Alene area. Bunker Hill & Sullivan was shipping out only handpicked, high-grade galena.

Skindiving for Old Bottles

Lake Coeur d'Alene, lying as beautiful as ever in its mountain setting of tightly braided evergreen trees, has naturally been the scene of other diving adventures. Phillip J. Gatlin of Moscow, Idaho, and his friends Bill Jones and Jim Howells, skindivers all, followed leads in my book *Steamboats in the Timber* to search Powderhorn Bay for some wrecked barges, Sanders Beach for traces of the *Miss Spokane,* and the steamship graveyard at Three Mile Point for hulls of old sunken boats. Even though making no discoveries of note they retained their enthusiasm, because as Gatlin puts it, "with each dive we find ourselves in a new world, unexplored, every inch covered with history."

On another trip they went up the St. Joe River, a main tributary to the lake, and explored a sunken tug some other divers had located. On upriver they investigated the site of the old waterfront town of Ferrell (completely gone now) and may have been the first to dive at what used to be the head of navigation in steamboat days.

Not far from where the river swings in from the steep north hillside which confines the valley, Phil and his friends found some remnants of pilings, and making descents along the nearby bank began to surface with all sorts of items: plates with blue borders, silverware, axes, a spittoon, pieces of iron chain, a brass bedstead with brass knobs, and a silver punch ladle inscribed, "Compliments of the Season — 1906 — Kreilsheimer Bros. Seattle USA." (How many cups of Tom and Jerry must have been poured from that ladle! — and how many stories of old time Ferrell it must represent!)

Bill Jones' mother, Mrs. William (Berny) Jones, Jr., of Calder, Idaho, a town located on up the St. Joe River, is a collector of antique bottles and she had alerted the boys to keep an eye out for interesting bottles. On one of the first dives Phillip Gatlin brought up a brown bottle, read the inscription, "Dr. Hostetter's Stomach Bitters," let out a laugh, and as he tossed the bottle back into deep water, said to Bill Jones, "Your mother'd never be interested in that one."

Bill, who knew that was just the one she WAS interested in, turned slightly green.

To understand Phil's big mistake you need to understand the whole mystique of bottle collecting. This is a hobby which has mushroomed so fast in the past decade that an estimated million Americans are now mad for bottles. A new kind of bottle club has come into being: an organization of people banded together to identify, date and price the antique glass containers they have dredged up. And since in the west many are dug from the refuse of old ghost towns, bottles are definitely a new kind of buried treasure. Interest and demand naturally spike prices, so the old saw about antiques in general now applies to bottles too: "If you think this is junk, just price it."

Bottle collecting has its phases and fashions as do other hobbies and at present among the most sought-after bottles are the old bitters bottles. Bitters had a high alcoholic content and was swigged down for "medicinal purposes" by all kinds of people who denied that intoxicants ever passed their lips. So from about 1850 to 1906 (when Pure Food laws spoiled the fun) the consumption of bitters was phenomenal, and collectors are now interested in getting good representations of all the different bitters bottles. Bill's mother very much wanted the Dr. Hostetter's bottle Phil had unknowingly tossed back into the river. The boys made any number of dives trying to retrieve it but had no luck.

Three years later Berny Jones went with the trio when they again dived at Ferrell. "The boys say the river is about forty feet deep there," Mrs. Jones writes. "When I noted a sudden explosion of air bubbles heading for shore I was uneasy. I felt something was terribly wrong. Then I saw a brown bottle and Bill's arm coming out of the water — and I knew in a flash he'd found the Dr. Hostetter's Bitters bottle. I love *this* bottle. It may not actually be worth more than $10 but I wouldn't part with it for $100."

In their various dives the boys also brought up whiskey bottles, beer bottles, wine bottles, household and cosmetic jars. One brown whiskey bottle with a bulbous neck was used by the Jimmy Durkin Company of Spokane about 1900.

So even if no one ever finds that lost silver ore there are old bottles aplenty to dive for in the Davy Jones' Locker of Lake Coeur d'Alene and its tributary rivers.

ON THE TRAIL OF DIAMOND DICK

THE POST FALLS DIAMOND MINE (Hult, *Lost Mines,* pages 193-195).

Richard Owens, who in 1907 claimed to have discovered a diamond mine on his farm five miles south of Post Falls, Idaho, was always afterwards known as Diamond Dick. Some old timers insisted that his gems, both in the natural state and when cut, looked like real diamonds, others believed them to be fakes, and still others thought he "salted" his mine. He seems to have had financial backing while he dug out the first pocket of stones. After that he was unable to raise money for further development work. When he died in 1946 it was still unsettled as to whether his diamonds were real or not.

Sworn Testimony

This is a case in which truth – or at least what was once sworn to as truth – has proven far stranger and more intriguing than legend.

Through the courtesy of Samuel May of Harrison, Idaho, a retired Army colonel who is the present owner of the Diamond Mine, I have been able to study a book-length typescript which brings the man known as Diamond Dick back to living, breathing reality and lets us hear in his own words his statements about his mining property.

The document is a 146-page transcript of testimony given at a U. S. Land Office hearing in Coeur d'Alene, Idaho, in July 1919. Immediately it establishes two facts which differ from the legend. The name of the principal is W. J. Owen, not Richard Owens, and his mining property lay not five miles south of Post Falls but less than a mile southwest at its closest corner.

According to information in the typescript Owen first came to Post Falls in 1905. By 1911 he had staked six quartz claims, the S.M.C., Rose Gold, Left Wing, Copper Queen, PayWell, and Truck Patch, and had bought an interest in a seventh, the Center Ridge, all in Section 9, T50N, R5W. He also staked a placer claim, the Kootenai, not far from the Spokane River in

Section 4. All eight claims were contiguous and comprised approximately 158 acres of land – that is, just under a quarter section. Owen lived on the place in a two-room house.

In 1914 he sold a half-interest in the claims to Charles Wendler, a farmer of Twin Lakes, Idaho, who invested $1200 in opening up the property. The development consisted of two long open cuts, one drift tunnel of twenty-six feet and two of thirty-six feet, one shaft of sixteen feet and another of twelve, and numerous shorter tunnels and prospect holes.

When Owen fenced the claims and ordered John Randall, a neighbor owning an adjoining farm, to keep his cattle off the property (even threatening "a little hot lead" if his warnings went unheeded), Randall retaliated by filing a homestead claim on the tract comprising the mining property, asking that it be granted him for pasturage purposes. Owen and his partner contested the homestead filing, insisting that the land was far more valuable for the minerals and gem stones found on it than for agriculture and accusing Randall of maneuvering to steal their rich mining claims.*

At the hearing before the Land Register and Receiver, Plaintiffs Owen and Wendler were represented by Attorney N. D. Wernette, and Defendant Randall by Attorney E. H. Berg. Witnesses were sworn and subjected to both direct and cross examination.

Owen was the first witness, his testimony soon establishing him as a quite different personality from the dreamy, eccentric and even slightly ridiculous Diamond Dick of legend. Here Owen seems a man who knows what he is about. At all times he sounds incisive and forceful; his replies are direct, exact and detailed.

After questions which elicited descriptions of the eight claims and the development work done on each, the following exchange took place:

---

* Legal description of the contested property:
NE 1/4 of NW 1/4, NE 1/4 of NE 1/4, NW 1/4 of NE 1/4,
Section 9; and Lot 13, Section 4 – T50N, R5W.
This includes three full forties plus a "lot" that is almost a forty, being thirty-eight acres, hence the total of 158 acres.

Q Now Mr. Owen did you find any mineral or ore on any of these prospects?
A Yes, sir.
Q On each one of these claims did you find it?
A On all of them.
Q Did you find any ore in any of these prospects in a defined ledge?
A Yes, sir.
Q There was ore in place there in the ledges?
A Yes, sir, I think so.
Q What kind of ore did you find?
A Gold, silver, copper, platinum, white and golden sapphires, rubies and garnets.

Owen then proceeded to offer a total of thirty-one different exhibits for the record. Nineteen were assay reports of mineral ores and placer gravels from the claims. He said little about the placer property except that his flumes and workings had recently been washed away by high water, but stated that the different assay reports of ores from the quartz claims showed values in gold, silver, and copper averaging from $28 to $80 a ton, and in one case of $900 a ton.

The other dozen exhibits were all of minerals taken from the quartz claims: "cloride of gold and cloride of platinum"; carbonate ore; ores allegedly containing gold, copper and silver; copper and gold bullion; and red stones and white stones Owen called sapphires and rubies – these being offered both in a rough state and in a finished form and accompanied by samples of the material in which the stones were found imbedded, some of the red stones still being in place.

The best of the red stones as well as the most promising of the mineral ores came from the Rose Gold claim. There, said Owen, the rubies appeared in surface outcroppings and in an open cut to a depth of thirty feet, being found in size from a pin point to twenty carats, and in "unlimited quantities." The red stones were also in evidence on two other claims, the Truck Patch and Kootenai, but development there had not progressed far enough for him to estimate their quantity. The white stones came from the Rose Gold and Truck Patch claims and were found in pockets of five or more, up to 150 carats a pocket.

In all he had taken about 150,000 carats of gem quality stones from the two claims but in 1915 had the misfortune to lose 100,000 carats along with numerous additional assay reports when his grip had been stolen from the Union Station in Chicago.

When questioned about the jewelers who had appraised his stones, what firms had done the cutting and to whom he had sold the finished product he showed no hesitation but responded not only with names but addresses. Two of the cutters were Chas. Hartman, 41-43 Maiden Lane, and L. Schula, 12 John Street, both New York City. He also had gems cut in Denver and Chicago. He had sold a small lot of rubies to the Jackson Jewelry Company in Chicago. And then he ticked off the names and in most cases the addresses of people to whom he had sold rubies or sapphires or both. A number of them lived in Spokane, one in Ohio, and some in Georgia (which was apparently Owen's home state). J. C. Howe of Spokane had worn one of the rubies "for six years and then sold it for $75." S. A. Wells, a mining man from Spokane who was to testify for Owen later in the hearing, had also bought and worn the gem stones.

It cost from $1.00 to $1.50 a carat to have the stones finished, said Owen; he could sell them for $6.00 to $10.00 a carat. But he had never sold enough to pay for his work, had he? — asked Attorney Berg in cross-examination.

Owen's answer was earnest if ungrammatical: "Well, you must figure, Mr. Berg, that all business for the last three years has been very quiet; only the paying mines that were in progress before the war has been dividend payers since the war came up, and the stone proposition had been on the quiet for the last four years."

And he insisted that a prudent man would be justified in working his mining claims.

Charles Wendler's testimony merely corroborated that of Owen's. Two Post Falls neighbors, W. H. Thomas and W. H. Sanders, also swore that Owen was prudent in working the claims. Thomas, when asked what kind of rock appeared in the ledges, answered, "Well, it's quartz bearing rock, I should judge would carry gold and silver, and the copper stained rock speaks for itself." He insisted the land was no good for pasturage, "Because two cows in there would starve to death. I have got only one cow and I wouldn't trust her in there."

And Sanders answered as follows:

Q You saw the rock with these rubies?
A Yes, sir.
Q And the other stones?
A Yes, sir, I saw the rock.
Q You took out some yourself?
A I took out some myself, had them finished.
Q The rock contains great quantities of them, does it?
A Yes, sir.
Q What is the fact as to whether or not there are well-defined ledges?
A Yes, sir, there are several well-defined for surface workings.

The outstanding witness for the plaintiffs was S. A. Wells. He was a lawyer with an interest in mining properties for thirty years. At present he was Secretary and General Manager of the Nickel Plate Mining Company, which was opening a mine in the Priest Lake area. Owen's gem stones had first been brought to his attention in 1913 by his son's father-in-law, a seasoned miner of fifty years' experience, part of it in the Kimberley mines of South Africa, who said that the formation on Owen's claims "was nearer like the formation of the Kimberley mines than any he had seen." Wells had then visited the Owen property, picked out a stone and bought it, the old miner in his family pronouncing it "a pure gem." When he lost this stone he bought three others from Owen. He had a white and a red one finished by Henry Lindemann of Denver and the red one mounted in a ring for his wife. A jeweler in Spokane had identified one stone as a sapphire and the other as a ruby and "said they were both good stones of that kind."

Wells had recently gone over the Owens claims and he testified that the quartz formation along the open cut on the Rose Gold claim showed great quantities of rubies. Pointing to an exhibit on the table, he said, "There is some of the rock right there and it is thickly studded with rubies; you can see it all through the formation on the face and sides of the open cut."

He mentioned one place "as large as that table I don't believe you could have laid a nickel down that you wouldn't have touched a little ruby from the size of a pin head up to larger sizes, just scattered on the surface." Yes, a reasonably prudent man would be justified in spending money to develop the claims; the Rose Gold claim could be mined for the rubies alone.

Then it was the turn for Defendant Randall and his witnesses to testify and one would not have known they were talking about the same piece of property. Randall and two of his neighbors, L. V. Lyon and John Knobel, all said there was no quartz on the claims, no veins, no ledges, maybe a little iron stain, but that meant nothing; and from a leading question put by Attorney Berg it was intimated that any copper showing in Owen's bullion might come from scrap copper thrown in during the melting process. Owen was accused of being more interested in logging and cutting cordwood on his claims than in mining, though even the statements of defense witnesses make it appear that most of the logging was done before Owen's time, and Owen himself swore he took out only those trees in his way and only cut already fallen timber for cordwood.

Randall and his neighbors admitted that the contested land was extremely rough, rugged and rocky but insisted it contained excellent browse for cattle and would support about eight milk cows the year around and thirty-five or more during the spring season.

The defense did not deny the prevalence of Owen's so-called gem stones, only tried to prove they had no commercial value. Lyon said he had taken a sample of similar stones to Boston where he had been told they were garnets and worthless, but he hedged in identifying all of Owen's red stones as garnets. Shown one exhibit he said, "I don't think I found that, I don't know what you would call that." Randall admitted that on the Owen property he had seen "something in the shape of the ruby that's in the rock here. That is the kind of rock I sent away in my assay."

The assay report referred to was the only one submitted by the defense, allegedly made from a "fair sample" of ore from the Center Ridge, even though this claim was not one on which

Owen himself had found any rubies or any especially rich ore. Randall said he told the assay firm (members of which he must have known) what the assay was for and that his reputation was at stake. The report showed the ore virtually valueless.

The most commanding defense witness was Neil Coventry, a civil and mining engineer who held the post of Kootenai County Surveyor and was in the process of becoming deputy U. S. Mineral Surveyor. He said there was nothing but granite on Owen's claims, no ledges or veins of minerals, and he believed that a prudent man would do best to pass the property by.

Though Coventry identified the red stones in one of Owen's exhibits as garnet, under cross-examination he admitted of other exhibits that he couldn't say what they were without a chemical test, and apparently in regard to some of the finished stones said, "I don't know about the others, they have been dressed up."

Such are the facts and contradictions found in the transcript. It should be noted that Owen made no claim whatsoever of having found diamonds nor did his detractors accuse him of trying to represent any of his stones as diamonds. Sapphires and rubies he swore he had but never any diamonds.

Owen and his witnesses certainly seem to have had the best of the argument, especially since the plaintiffs introduced nineteen assay reports as against one by the defense and twelve mineral exhibits as against two. But either the arguments and exhibits were not so impressive and convincing as they sound or else there was local hanky-panky, for the Register decided against Owen and Wendler and approved the Randall homestead application. But Owen and his partner fought on. They appealed the case to the General Land Office in Washington, D.C., where the final decision was rendered in Owen's favor.

After that the whole thing reverts to legend — the legend of Diamond Dick. Did Owen later decide to call his white stones diamonds? Or did the idea of gem stones suggest diamonds to most people so that they used the Diamond Dick nickname in jest?

That Owen in later years did have stones he called diamonds is borne out by a letter from John E. Campbell of Opportunity, Washington. Campbell and a friend, Don McLean, both being rockhounds and adventure enthusiasts, decided to follow up on the Diamond Mine story and try to learn if Dick had found anything of value. They inquired around Post Falls and located an old friend and neighbor of Dick's.

"He had five of the diamonds," writes Campbell, "plus a garnet cut into sets, which he had obtained before Dick died. The so-called diamonds seemed to lack the sparkle of real diamonds, so we bought one from the man and took it to a jeweler friend for an appraisal. The jeweler pronounced it a fairly well cut piece of clear quartz. For good clear quartz can be cut into a rather handsome looking stone, though if put beside a real diamond it immediately shows its deficiencies. Such quartz stones are worthless."

Dick's friend took Campbell and McLean to the site of Owen's former mining activity, where two of the old tunnels still remained. "We found a lot of clear quartz," writes Campbell, "and a lot of small garnets, from which we deduce that the garnet and quartz 'diamonds' came from the mine . . . . Our personal evaluation of the colorful character, Diamond Dick, is that he found the quartz crystals and sent them back east (and) was then informed they were not worth cutting. But rather than to back down on his fabulous claims, and possibly hoping even to sell some of his gems, he kept talking to all who would listen about his 'diamonds' — and was always trying to get a grubstake so he could make a new strike."

This theory would make Owen out a lying fraud, which perhaps he was. But in the transcript he sounds a convinced and dedicated miner, and it seems odd that a fraud should have been so direct and non-evasive, should have given so many names and addresses that could be checked, and should have produced so many assay reports and mineral exhibits. It seems even stranger that a fraud could get four other people to back up his delusions with sworn statements. In particular S.A. Wells seems a man of high caliber and his story about the old Kimberley miner's enthusiasm for Owen's stones could hardly have been made up.

After Owen and Wendler won the appeal in Washington, D. C., Wendler bowed out as a partner. He probably had paid all the costs in the case and found he had spent more than he could afford. Afterwards Owen had trouble in finding other investors, though from the legend he seems to have clung tenaciously to hope. But at length being unable to keep up the mining assessment work he was forced to let the land revert to the government.

When the tract was put up for sale by the Bureau of Land Management, Samuel May, who had already acquired the adjoining Randall homestead, bid it in. Only later did May learn about the past mining activity on the property and obtain a copy of the transcript. He is unconvinced by all the statements about sapphires and rubies made at the hearing. He considers the old Owen mine worthless and says he has no intention of spending money to prove the contrary. The one tunnel which erosion has not claimed is so dangerous to youngsters he is planning to dynamite it.

So in the end we remain as baffled by old Dick and his "diamonds" as in the beginning. Only one thing is certain: If Owen's white gem stones were only clear quartz and his red stones only garnet, then the Coeur d'Alene Land Register in July 1919 heard five different people commit some high and mighty perjury.

"UNCLE BILLY" WAS A MEMORABLE CHARACTER

INDIAN POST OFFICE MINE (Hult, *Lost Mines,* pages 199-203).

On a ridge along the Lolo Trail, on the divide between the Locksa River and the North Fork of the Clearwater, a certain cairn of stones was used by the Indians as a place to leave stick and rock messages. This cairn, which still stands, is known as the "Indian Post Office," and has served as a guidepost to an Indian lost mine, supposed to lie a half-day's journey beyond — possibly near Cayuse Junction. An old prospector, Bill Bartlett, sought the mine for many years and once had a vivid dream pointing the way. He found his dream campsite on Cayuse Creek but failed to locate the gold.

Since legend credits the Nez Perce Indians with using the Indian Post Office as a landmark in journeying to a place where they obtained gold, it is of interest that Captain E. D. Pierce, the white man who first discovered gold on a tributary to the Clearwater River, not only obtained his intelligence from members of the Nez Perce tribe but was led to the right stream by a Nez Perce girl guide.

On the Subject of the Clearwater Country

Letter of comment from George K. Oliver, Hood River, Oregon:

"I was born and raised in the Orofino, Idaho, country on the Clearwater River above Lewiston. Your reference to Charlie Lame and Robert Bailey of Lewiston brought back memories of these very fine gentlemen with whom I was once very well acquainted.

"I also have many remembrances of 'Uncle Billy' Bartlett, whom you mention in connection with the Indian Post Office Mine.

"About 1915 my brother-in-law and I had been up on the state land eight or ten miles east of Orofino cutting cedar shake bolts. We had our load on a lumber wagon and were returning to Hell's Daylight (now known by the tamer name of Harmony Heights) over the old ridge road that connects Orofino with Pierce via Red's Prairie. Uncle Billy rode up behind in a typical loose jointed western fashion, tied on and climbed aboard to visit the few miles to our turn-off.

"I suppose I had seen him before but for some reason the impression of that day has remained in my mind. He was angular of frame and general appearance, standing about five feet ten and weighing no more than 150 pounds. He was then, as ever, deeply tanned, with steel blue eyes and a high thin nose. He was dressed in low-heeled boots, dark woollen trousers, faded red shirt, dark vest, with a blue bandanna at his throat. A black slouch hat topped a heavy head of salt-and-pepper hair, General Grant beard and mustache. He spoke in a rather high voice with a touch of New England or down East in it.

"I never heard him use profanity but imagine him capable of it. And though on that hot afternoon he expressed a wish for a few beers, it appears this was mere nostalgia. For after never drawing a sober breath for forty years he had turned to the Bible and no longer imbibed. Of this fact I have been reminded by my sister Blanche Casteel of Orofino. She knew Uncle Billy for years after I left that part of the country.

"He used our homestead on the breaks of Orofino Creek as a regular stopping place on his annual visits outside from his Pike's Creek diggings, located on a branch of Poorman Creek, a small stream which flows into Orofino Creek. He was on old-time fiddler of no small merit and could be relied on to play for his meals.

"While on the subject of the Clearwater country I would like to call your attention to one small error. You mention the discovery of gold on Orofino Creek, a tributary to the North Fork. It is true that some of the Pierce City Mining District is drained by the North Fork, namely Reed's Creek, Alder Creek, Gold Creek, Snake Creek, Oro Grande and French Creek areas. These were for the most part only prospect and 'pocket' diggings. Whereas the main wealth of 'rich old Pierce City' lay in Canal Gulch, Rhodes Creek, Quartz Creek and in later years Cow Creek, all of which drain into Orofino Creek, which empties into the main Clearwater at my home town of Orofino. Ahsahka, at the mouth of the North Fork, is four miles farther down the river . . . . Just thought you might wish to correct this unimportant bit of misinformation.

"I also know a little other treasure lore which might interest you. I had an uncle who spent a summer or two looking for the Lost Robinson Mine. The story goes something like this:

"During the Pierce City years Robinson would appear in Lewiston with plenty of dust. In the spring he would be followed up the North Fork by those envious of his wealth, but twenty-five or thirty miles above the mouth of this river, always near the mouth of Cranberry Creek he would disappear, eluding his pursuers. But come fall he would reappear, heavy with dust and nuggets.

"After several seasons he just failed to show up again. Was it a pocket he worked out? A crevice somewhere along the North

Fork? Or was he simply using this route as a backdoor into the Pierce country? No one can say.

"Here's another little mystery. While I was working at the Bungalow Ranger Station on the Clearwater one fall in the middle 1920s, I met an aged man who had come to that point, which was the end of the road. With him were two vigorous, intelligent young men whom he sent down the river trail on foot, carrying heavy packs. After the boys had been gone several days and he appeared confident they would find what they were looking for he confided in me that they were his nephews, and he proceeded to show me a piece of what appeared to be very rich gold quartz. He told me how years ago, before there were any trails or roads in that area he had been prospecting on Isabelle* Creek and had found this rich quartz vein behind a thin curtain of water veiling a little falls about four or five miles above the mouth of the creek. He had gone out expecting to return, but due to ill health in the family and other adverse fortunes he had been forced to delay many years. He described the exact location to me, which I later visited. I found the falls but not a sign of quartz.

"When his two nephews reported back footsore, hungry and weary, with no gold, he changed before my eyes from an alert old gentleman to a broken old man. It was sad — and yet he had for two weeks cherished a dream of hope that seldom comes to a man past eighty, and the two boys had had a trip they would not soon forget.

"Lately I've been wondering if those boys looked in the right place on Isabelle Creek. I am only just sixty years of age. My packsack hangs in the shed, and — say — where are my gold pan and quartz hammer?"**

---

* Maps of the area use the spelling Isabella.

** Before he had a chance to take down his packsack and head for the Clearwater country George K. Oliver died in Hood River on April 4, 1969.

A TRUE PREDICTION

WHITE BIRD, BLACK TAIL (Hult, *Lost Mines,* pages 223-224)

According to legend, two caches are buried somewhere near White Bird, a town in the Salmon River country of Idaho. One consists of "lots of gold" hidden by an Indian named Black Tail. The other is a hoard of $75,000 in gold dust stolen from miners en route from Florence to Lewiston. The robbers, being pursued, secreted the loot among the rocks along the Salmon River. Later they were all killed.

A woman who appeared in White Bird in 1903 to hunt the Black Tail cache claimed to be a spiritualist medium and to have foretold the Heppner, Oregon, flood which earlier that year on June 14 had swept down the dry gulch and drowned 200 people.

Though the woman and an old man who was helping her dug a good many holes, it is believed that her psychic powers deserted her and she was unable to locate the treasure.

A Prophecy Fulfilled

Letter of comment from Duane Alderman, Milton-Freewater, Oregon:

"I am originally from Heppner, Oregon. When I was eighteen years old I found an old newspaper in a homestead, and I have kept it all these years. It is the Heppner *Gazette-Times,* dated June 25, 1903."

And he quotes the following excerpt from that paper:

"On Friday before the storm on Sunday, two very old squaws, wrinkled and weatherbeaten from the storms of many moons, stopped at the warehouse of the Morrow County Land & Trust Company to sell a small consignment of pulled wool. As the weather was threatening, R. F. Hynd, the manager of the warehouse, spoke to the squaws and told them they had better go or they would get wet.

"The older squaw got up and said: 'Yes, we go to the hills, heap high. You look out. Not long great water come take Bos-

ton man, papoose, horse, houses and everything to the Columbia.'

"The *prediction* was looked upon lightly and furnished no little amusement. BUT THE PROPHECY WAS FULFILLED, EVEN IF IT DID COME FROM AN INDIAN."

Alderman's letter continues, "So the statement of the spiritualist medium in the White Bird story must be true."

The old newspaper account does help to confirm at least one of the facts about the 1903 search for Black Tail's cache. And perhaps the Indian woman with the rare prophetic gift may have had better luck in her quest than skeptical white onlookers believed.

And so

THE END

and to all –

GOOD HUNTING!

ACKNOWLEDGMENTS

In connection with the special services they represent I am indebted to Verda K. Hansberry, Head, Non-Resident Loans, Public Library, Seattle, Washington, and Mildred A. Cline, Head, Readers Services Division, State Library, Salem, Oregon.

I also wish to thank the following librarians for thoughtfulness and diligence in answering my many queries and helping to solve knotty research problems: Robert D. Monroe, Curator, Special Collections, and Mildred K. Sherwood, Northwest Collection, University of Washington Libraries, Seattle; Hazel E. Mills, Washington-Northwest Room, Washington State Library, Olympia; and Elizabeth Ann Johnson, Head, Literature and History Department, Public Library, Portland, Oregon.

I also appreciate the research assistance given by Florence K. Palmer, Portland, Oregon, and John D. McGough, Kent, Washington.

BIBLIOGRAPHY

LOST VAULT OF GOLD

Sam L. Simpson, "The Lost Cabin," *Oregon Native Son,* Vol. 2, pp. 202-208.

Sam L. Simpson, poem, "Beautiful Willamette," *Oregon Native Son,* Vol. 1.

W. W. Fidler, "Personal Reminiscences of Samuel L. Simpson," *Oregon Historical Quarterly,* Vol. XV, pp. 264-276.

Ray H. Glassley, article about Sam L. Simpson, Portland *Oregon Journal,* January 8, 1956, magazine section, p. 12.

Ray H. Glassley, *Pacific Northwest Indian Wars.* Portland, Oregon: Binfords & Mort, 1953.

H. H. Bancroft, *History of Oregon.* 2 volumes. San Francisco: History Company, 1886. Vol. II, chapters on Rogue River Indian wars.

The Golden Frontier, The Recollections of Herman Francis Reinhart, 1851-1869. Edited by Doyce B. Nunus, Jr. Austin: University of Texas Press, 1962.

INDIAN LOST CINNABAR MINE

"Cinnabar in Southern Oregon — The Unfound Rich Ledge," *West Shore* (Portland, Oregon), September 1875, pp. 1-2.

Untitled short item, *West Shore,* November 1878, p. 73.

H. H. Bancroft, *History of Oregon.* Vol. II, chapters on Rogue River Indian wars; also "Mines and Mining," pp. 742-743.

Langley's San Francisco Business Directory, for years 1859 and 1860. (Available in the San Francisco Public Library).

A. G. Walling, *History of Southern Oregon.* Portland, Oregon: A.G. Walling, 1884.

A. E. Kellogg, "Cinnabar Ore in Jackson County," *Engineering and Mining Journal,* May 3, 1919, p. 787.

Omar C. Spencer, *The Story of Sauvies Island.* Portland, Oregon: Binfords & Mort, 1940.

Notations relative to the history of the Jewett family, pioneer file, Oregon Historical Society, and newspaper index, Public Library, Portland, Oregon.

Walter W. Bradley, *Quicksilver Resources of California.* California Mining Bureau Bulletin No. 78 (1918).

Up and Down California in 1860-1864. The Journal of William H. Brewer. Edited by Francis P. Farquhar. Berkeley and Los Angeles: University of California Press, 1966. Chapter II, New Idria; Chapter III, New Almaden.

Hoover and Rensch, *Historic Spots in California.* Stanford: Stanford University Press, 1964. The New Idria Mine, p. 288; New Almaden, pp. 329-30.

ED SCHIEFFELIN'S LOST OREGON GOLD MINE

Manuscript, "Edward Schieffelin, The Discoverer of Tombstone, Arizona," six pages; "History of the Discovery of Tombstone, Arizona, as told by the Discoverer, Edward Schieffelin," fifteen pages. Copy reproduced from the original in the Arizona Pioneers' Historical Society Library, Tucson, Arizona.

County records, Jackson County, Medford, Oregon; Douglas County, Roseburg, Oregon.

Stewart H. Holbrook, "Ex-Oregon Youth Real 'He-Man of the West,'" Portland *Oregonian,* September 4, 1932, magazine section, p. 5.

Stewart H. Holbrook, "The Guy Who Named Tombstone," *American Mercury,* June 1944.

Dale Vincent, "Lost: Ed's Red Blanket and Rich Gold Mine," Portland *Oregonian,* March 31, 1946, magazine section, p. 8.

"Missing Red Blanket May Be Clue to Ed Schieffelin's Lost Mine in Canyon Creek Region," Roseburg, Oregon, *News-Review,* April 15, 1948.

J. Frank Dobie, "Schieffelin's Gold," *True West,* October 1958, p. 13.

Carl Gohas, "Schieffelin Struck It Again, But Where's His Red Blanket Mine?" Portland *Reporter,* October 23, 1963, p. 11.

"Ed Schieffelin's Lonely Death," Ashland, Oregon, *Tidings,* May 17, 1897.

Walter Noble Burns, *Tombstone, an Iliad of the Southwest.* New York: Doubleday, Doran & Company, 1927.

Ellis Lucia, "Oregon's Liveliest Ghost Town," *Ford Times,* August 1952, p. 25.

Jacksonville, a pictorial review – a living town of the old west. Western trails publications, Box 508, Medford, Oregon.

R. W. McAllister, *Lost Mines of California and the Southwest.* Thomas Brothers, 2560 Glendale Blvd., Los Angeles, and 2308 Market St., San Francisco, 1953.

Oregon's Historical Old Jacksonville. Mr. and Mrs. Robert N. Beyer, publishers, 707 N. Oregon Street, Jacksonville, Oregon.

Francis D. Haines, Jr., *Jacksonville, Biography of a Gold Camp.* Available at Jacksonville Museum or Oregon Historical Society. Copyright 1967.

Jacksonville Museum leaflets.

COXCOMB HILL CACHE, FLAVEL CACHE, FISHHAWK FALLS AND OTHER LOST MINES OF CLATSOP COUNTY

Charles R. Eastland, Astoria, Oregon. Narrative accounts of lost mine legends; correspondence.

Emma Gene Miller, *Clatsop County, Oregon, Its History, Legends and Industry.* Portland, Oregon: Binfords & Mort, 1958.

Historic Astoria, Oregon, and *Clatsop County, Oregon,* folders, Astoria Chamber of Commerce.

Fort Clatsop National Memorial, folder, National Park Service, U.S. Department of Interior.

Charles Nordhoff, *Northern California, Oregon and the Sandwich Islands.* New York: Harper & Brothers, 1874.

Helen Hunt Jackson, *Glimpses of Three Coasts.* Boston: Roberts Brothers, 1886.

Virginia Jensen, "Mining Claims Scanty in Clatsop County – Tree to Tree Description Given for First Claim," *Astorian Budget,* March 16, 1953.

"Columbia May Become Source of Revenue Through Minerals," *Daily Astorian,* August 15, 1961.

Burnby M. Bell, Astoria, Oregon. Correspondence.

The Bunker Hill Company, Kellogg, Idaho. Correspondence.

LOST GOLD STRIKES OF THE OLYMPIC MOUNTAINS

E. N. Richardson, Tonopah, Nevada, "Hunter Tells Amazing Yarn – Discovery of Gold is Told by Aged Nevada Prospector," and "Desert Rat Tells about Wanderings in Olympics," no dates or source, Costello Scrapbook No. 2, p. 22 (available in the University of Washington Northwest Collection).

Ruby El Hult, *Untamed Olympics, The Story of a Peninsula.* Portland, Oregon: Binfords & Mort, 1954.

Mary Gay Morse, *Lore of Olympic-Land.* Los Angeles, California: Biola Press, 1924.

LOST MINES OF THE MOUNT ADAMS AREA

"Hunting for a Lost Gold Mine – Ranchers Organize Company to Discover where Rich Quartz Came From," Seattle *Post-Intelligencer,* August 9, 1909, p. 4, c. 6. Dateline Husum, Aug. 4 (Special).

Washington: A Guide to the Evergreen State. Works Project Administration, 1941.

Mt. Adams Wilderness – Gifford Pinchot National Forest. U.S. Department of Agriculture – Forest Service, Pacific Northwest Region. Map and legend, 1965.

C. E. Rusk, *Tales of a Western Mountaineer.* Boston and New York: Houghton Mifflin Company, 1924.

News item, Seattle *Times,* October 8, 1937, from Portland, Oregon.

CHIEF KITSAP'S GOLD MINE

"A. P. Carr of Tacoma Thinks He Has Found the Treasure," Seattle *Post-Intelligencer,* July 28, 1900, p. 3. c. 3.

"Prominent Pioneer Passes," and "Leading Pioneer of City Passes – Anthony Carr, Who Settled Here in 1865, Taken by Death," clipping files, Washington Historical Society. Tacoma newspapers, no dates but apparently April 1923.

"The Indian, Chief Kitsap," *Pacific Northwest Quarterly,* Vol. 25, pp. 297-301.

"Van Ogle's Memory of Pioneer Days," *Washington Historical Quarterly,* Vol. 13, p. 275.

Item about Chief Kitsap, *Washington Historical Quarterly,* Vol. 1, No. 2, p. 6.

LEGENDARY LOST GOLD MINE OF THE COEUR d'ALENES

"Lost Cabin Mine Said to Have Been Rediscovered by a Woman," Portland *Oregonian,* September 20, 1896, p. 5. (Apparently picked up from the Spokane *Chronicle*.)

An Illustrated History of North Idaho. Western Publishing Company, 1903.

Edmund R. Cody, *History of the Coeur d'Alene Mission of the Sacred Heart.* Caldwell, Idaho: Caxton Printers, Ltd., 1930.

C. S. Kingston, "Coeur d'Alene Gold Excitement of 1864," *Review,* a publication of the Eastern Washington College of Education, Cheney, Spring, 1942.

News item about gold strike on West Fork of Pine Creek, Kellogg *Evening News,* August 13, 1928: recently reprinted in that newspaper's column "In the Old Days."

Henry L. Reimers, Spokane, Washington; Zeke Older, formerly of Medimont. Correspondence.

"The New Mines," Portland *Oregonian,* June 6, 1865.

"Mineral Ridge: Yesterday – Today – Tomorrow" and "Mineral Ridge Scenic Area." Publications of Coeur d'Alene District, Bureau of Land Management, U.S. Department of the Interior.

POST FALLS – HUETTER CACHE

Preston E. Onstad, Department of English, Oregon State University, Corvallis, Oregon. Correspondence.

Thomas J. McAndrew, North Hollywood, California. Untitled manuscript of reminiscences (in collection of Ruby El Hult McAndrew).

Marceline Saltice Kevis, Worley, Idaho. Interview.

Hazel Corbeill, Kellogg, Idaho. Correspondence.

OUT FROM SANTA CREEK – SOME LOST MINE TRAILS

Al Vanderpoel, Los Gatos, California. Correspondence and interview.

Al Vanderpoel, "The Lost Mine That Wasn't Lost," an unpublished article; "Hardrock Johnson of the Buckhorn Mine," *Panning Gold, the Prospector's Guide,* issues Nos. 59 and 60, April-July, 1959; "Prospect That Old Diggings," *Panning Gold,* issue No. 71, April-May 1961. (NOTE: This interesting little magazine, official publication of United Prospectors, Inc., no longer appears).

Arthur B. Johnson, St. Maries, Idaho. Correspondence.

Bibliography

THE SPOTTED LOUIE MINE

Bill Smith. Correspondence.
Jim Estes, Clarkia, Idaho. Correspondence.
Mr. and Mrs. Charles W. Kevis, Worley, Idaho. Interview.
Ray L. Hilding, Supervisor, and Lawrence Glover, St. Joe National Forest, St. Maries, Idaho. Interview.
Orland A. Scott, *Pioneer Days on the Shadowy St. Joe.* Caldwell, Idaho: Caxton Printers, Ltd., 1967.
"Finding of Rich Gold Mine Told," Spokane *Spokesman-Review,* December 7, 1960 (dispatch from Wallace, Idaho).

AN OLD GREEN COPPER STILL

Lee Carpenter, Jr., St. Maries, Idaho. Correspondence.

CHINEE CACHE OF CHINA GULCH

Bill Smith. Correspondence.
Jean Davis, *Shallow Diggin's, Tales from Montana's Ghost Towns.* Caldwell, Idaho: Caxton Printers, Ltd., 1963. Chapter 25, "Cedar Creek, Louisville, Forest City, Mayville."

MINER'S BOOTFUL OF GOLD

Jean Davis, *Shallow Diggin's.* Chapter 23, "Cooke City."
Muriel Sibell Wolle, *Montana Pay Dirt; a Guide to the Mining Camps of the Treasure State.* Denver, Colorado: Sage Books, 1963.

NEAHKAHNIE TREASURE

F. L. Watrous, Forest Grove, Oregon. Interview and correspondence.
Ruth Gates, Assistant Curator, Washington County Historical Society, Hillsboro, Oregon. Letter of information about the Iler family.
Robert L. Benson, Map and Print Service, Hillsboro, "The Iler Family," one-page typescript.
Items about the Iler family, newspaper index, Public Library, Portland, Oregon.
Burford Wilkerson, Tillamook, Oregon. Correspondence and gift of photographs.
"The Oregon Beeswax Mystery," *Shell News*, December 1961, pp. 20-21.
Mrs. Norman P. (Betty) Allen, Portland, Oregon. Correspondence.
Mr. and Mrs. Harry Tuttle, Winthrop, Washington. Letters addressed to Alex Walker and Burford Wilkerson, Tillamook, Oregon.
"Spanish Treasure," Tillamook County Chamber of Commerce. Two mimeographed sheets dated January 15, 1962.
Mrs. Loran Lawrence, Tillamook, Oregon. Correspondence, photos and letters pertaining to Chinese coin.

"Unusual Find," letter to the editor from Howard Humphrey, Jr., Junction City, Oregon. Clipping, no date, no newspaper source.

Stan Allyn, "Huge Rusted Anchor Seen as Link to Legendary Treasure Ship," Portland *Oregonian,* December 4, 1961

Don Viles, Garibaldi, Oregon. Correspondence and gift of photographs.

"Tony Mareno's Treasure Hunt at Neahkahnie," scrapbook compiled by Ruby El Hult McAndrew from clippings sent by Stanley Partridge, Garibaldi; Al and Alice Reiner, Stayton; Eileen Best, Springfield; Mrs. Loran Lawrence, Tillamook – all of Oregon; Ron Gluth, Tacoma, and Jean Teig, Seattle, Washington; Hazel Corbeill, Kellogg, Idaho. News stories are from Tillamook *Headlight-Herald,* Eugene *Register-Guard,* Salem *Oregon Statesman,* Portland *Oregonian* and *Journal,* Spokane *Chronicle,* and Seattle *Post-Intelligencer.*

Ralph Gironda, Jr., New Rochelle, New York. Correspondence.

J. Frank Dobie, *Coronado's Children – Tales of Lost Mines and Buried Treasures of the Southwest.* Garden City, New York: Garden City Publishing Company, Inc. (Copyright 1930, The Southwest Press). Chapter XIX, Shadows and Symbols, pp. 333-341.

BLUE BUCKET MINE

Keith Clark and Lowell Tiller, *Terrible Trail: the Meek Cutoff, 1845.* Caldwell, Idaho: Caxton Printers, Ltd., 1966. Correspondence; interview with Lowell Tiller.

"Lost Emigrant Mines", Eugene *Oregon State Journal,* November 16, 1877, p. 4.

James Field Journal, 1845, "Crossing the Plains," Portland *Willamette Farmer,* April 18 through August 1, 1879.

John Herren, "A Diary of 1845," from an Albany, Oregon, newspaper. (Taken from a diary lost in a fire at Heppner, Oregon, these excerpts cover events along the trail from August 23 through September 8, 1845.)

Jacob Manley Currier, Corvallis, Oregon, letter dated September 2, 1919. Clipping in Personal Scrapbook No. 7, p. 8, Schminck's Museum, Lakeview, Oregon.

Ora M. Yeager, Castle Rock, Washington, "The Blue Bucket Mine," two pages, and "The Story of the Blue Bucket Mine," eight pages. Unpublished manuscripts. Also correspondence.

Leslie Shaw, "Blue Bucket Mine – Group Believes Fabled Site Found in Lake County," Lakeview *Lake County Examiner,* April 28, 1960.

Leslie Shaw, "The Blue Bucket Mine Story," Lakeview *Lake County Examiner,* May 5, 1960.

"The Blue Bucket Route – Book Details Route of 1845 Lost Wagon Train," Lakeview *Lake County Examiner,* October 25, 1962.

Lois A. Pierce, *Lost Immigrants of 1845 and the Blue Bucket of Gold.* Printed by the Shelton-Mason County Journal, Shelton, Washington, 1962.

Lancaster Pollard, "Scientific Safari Determines Where Blue Bucket Mine is Not," Portland *Oregonian,* November 24, 1958.
Choral Pepper, "The Case of the Blue Bucket Gold," *Desert,* September, 1967.
Jack Pepper, "The Oregon Desert," *Desert,* October, 1967.
Harry Sinclair Drago, *Lost Bonanzas.* New York: Dodd, Mead & Company, 1966. "Nevada's Blue Bucket Bonanza."
A. J. Splawn, *Ka mi akin—the Last Hero of the Yakimas.* Yakima, Washington: Mrs. Margaret C. Splawn, 1958.
R. W. Pitney, Junction City, Oregon. Correspondence.
H. H. Bancroft, *History of Washington, Idaho and Montana.* San Francisco: The History Company, Publishers, 1890.

THOSE LOST CABIN MINES

A. A. Reiner, Stayton, Oregon. Interview and correspondence. Also reminiscences set down by Mrs. Reiner: "The Story of Mr. Morton," "The Old Man with the Rocking Chair," "Mr. Hunsaker," and "Newspaper Correspondent."
Mr. and Mrs. K. C. Boedecker, Portland, Oregon (now of Salinas, Kansas). Correspondence.
Norman O. Rosenberger, Sr., Sandpoint, Idaho. Correspondence.
L. J. Edmunds, Cave Junction, Oregon. Correspondence.
Samuel T. Frear, Public Information Specialist, Willamette National Forest, Eugene, Oregon. Correspondence.
Metzger *Atlas of Klamath County.*

LOST TILLAMOOK MINE

William Perry, McMinnville, Oregon. Correspondence.
Don Viles, "Hidden Tillamook Nuggets," *The Buccaneer,* October, 1967.

PORTLAND TREASURE CHART

Mrs. Arthur House, Portland, Oregon. Telephone conversation and correspondence.
Ruth Burgess, Portland, Oregon (now of Los Gatos, California). Correspondence.
Ronald L. Gluth, Portland, Oregon (now of Tacoma, Washington). Correspondence.
Priscilla Knuth, Oregon Historical Society, Portland, Oregon. Correspondence.
Pioneer file, Oregon Historical Society, Portland.

IN THE LOST FOREST

Leslie Shaw, Editor, Lakeview *Lake County Examiner.* Correspondence.
"Gold Discovery Revives Legend," *Lake County Examiner,* March 12, 1925.

Don C. Tracy, Lakeview, Oregon. Correspondence and narrative account of his various uranium discoveries.

Clair D. Smith, Lakeview, Oregon. Correspondence and narrative account of the discovery of the Lucky Lass Mine.

"Uranium History in Lake County Dates from 1955 Discovery," Lakeview *Lake County Examiner,* March 10, 1966.

THE CAPTAINS' CACHES

Robert Nash, "Ilwaco's Gold Mystery," *Frontier Times,* December-January 1967, pp. 24-25. Also correspondence with the author.

Charles R. Eastland, Astoria, Oregon. Correspondence.

CHEVY CHASE TREASURE CHEST

"Treasure Mystery Revived – Woman Claims Gold Sovereign Is One of Coins Long Sought Here," Port Townsend *Leader,* October 1, 1959. Clipping sent by Mrs. Florence E. Hicks, Bremerton, Washington.

Richard F. McCurdy, The Leader Company, Port Townsend. Correspondence.

Philip W. Bailey, Chevy Chase, Port Townsend, Washington. Correspondence.

James G. McCurdy, *By Juan de Fuca's Strait.* Portland, Oregon: Binfords & Mort, 1937.

LOST SPANISH SILVER MINE

Leon A. Patten, Spokane, Washington. Correspondence.

John B. Thompson, Bonners Ferry, Idaho. Correspondence.

George H. Unternahrer, Bonners Ferry, Idaho. Correspondence.

Allan H. Smith, Chairman, Department of Anthropology, Washington State University, Pullman, Washington. Correspondence.

Report, District Forest Ranger to Forest Supervisor, Kaniksu National Forest, Sandpoint, Idaho, April 11, 1938.

"Search for 'Pat Shea Mine' " reveals Ancient Excavation, *The Sundial,* No. 8, p. 2.

Struthers Burt, *POWDER RIVER Let 'er Buck.* New York: Farrar & Rinehart, 1938.

Steve Wilson, "White Mountain's $10,000,000 Secret," *Argosy,* March 1966.

Keith Clark and Lowell Tiller, *Terrible Trail: The Meek Cutoff, 1845.* Caldwell, Idaho: Caxton, Printers Ltd., 1966. Statement of Flora E. Bailey, pp. 116-118, and footnote p. 117.

Merle W. Wells, Historian and Archivist, Idaho Historical Society, Boise, Idaho. Correspondence.

Copy, acquisition card, Object No. 1757, received June 16, 1949, Idaho Historical Society, Boise, Idaho.

James Hausam, Sedalia, Missouri and Joe Rasmussen, Spangle, Washington. Correspondence.

"Rocky Bar Spanish Legend," Reference Series No. 19, Idaho Historical Society, Boise, Idaho.
Arthur A. Jacobson, Enaville, Idaho. Interview.

SUNKEN ORE OF LAKE COEUR D'ALENE

Spencer Campbell and Jerry McFeeley, Paine Field, Everett, Washington. Interview.
"Lost Bunker Hill Silver Again Sought," Kellogg *Evening News* July 8, 1960.
"Divers Discover Lost Silver Reef," Spokane *Chronicle,* July 10, 1960.
"Lake Treasure May Stay There," Kellogg *Evening News,* July 11, 1960.
Ozzie Walch, Coeur d'Alene, Idaho. Correspondence.
Phillip L. Gatlin, Moscow, Idaho. Correspondence.
Mrs. William (Berny) Jones, Jr., Calder, Idaho. Correspondence.

POST FALLS DIAMOND MINE

Samuel May, Harrison, Idaho. Correspondence.
Typescript of legal testimony, U.S. Land Office hearing, Coeur d'Alene, Idaho, beginning July 21, 1919, 10 a.m. 146 pages.
John Campbell, Opportunity, Washington. Correspondence.

INDIAN POST OFFICE MINE

George K. Oliver, Hood River, Oregon. Correspondence.
Mrs. Blanche Casteel, Orofino, Idaho. Correspondence.

WHITE BIRD, BLACK TAIL

Duane Alderman, Milton-Freewater, Oregon. Correspondence.
Article from Heppner *Gazette-Times,* June 25, 1903.

INDEX

Index

DE

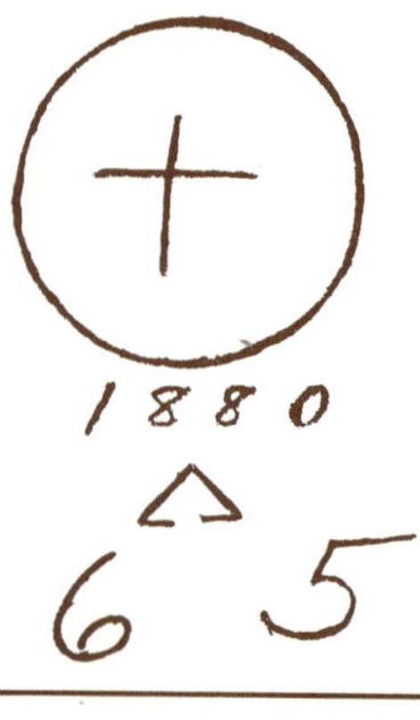
1880
6 5